DESIGN OF WOOD STRUCTURES

A PRIMER WORKBOOK

PYO-YOON HONG, PH.D., P.E.

Matthias & Alex Publishing, Inc.

Design of Wood Structures
A Primer Workbook
Pyo-Yoon Hong

Publisher's address and contact information:
376 Creek Manor Way, Suwanee, GA 30024

ISBN-10: 0989511200
ISBN-13: 978-0-9895112-0-9

Printed in the United States of America

Book Design by Pyo-Yoon Hong
Cover Photo by Sander Westerveld
Editing by Pyo-Yoon Hong

First Edition: June 6, 2013
10 9 8 7 6 5 4 3 2 1

This book is dedicated to

Yoo-Young

Seong-Un

Seong-Jun

I love you all, each and every one.

TABLE OF CONTENTS

Chapter 4

Chapter 5

Chapter 6

Chapter 7

Chapter 8

Chapter 9

Appendix A

Appendix B

PREFACE

This book offers a concise and thorough presentation of wood design process, application and underlying structural principles and thus is committed to developing users' problem-solving skills. This workbook makes the contents of textbooks with same subjects more visible, extractable, and relevant for an application or process. The material is reinforced with variety of structural design examples of progressively varying degrees of difficulty to illustrate structural principles and design issues that focus on practical and realistic situations encountered in professional practice. This book features many photorealistic figures that have often been depicted in 3-dimensional view to appeal to visual learners. The case study problems and group workshop are prepared to relate the verbal and visual elements to each other in an effective way. Most verbal elements are presented in categorized boxes. Some of the visual and verbal elements are deliberately left incomplete or missing so the instructor and students can complete them together in the classroom. This approach promotes problem-based learning and active participation of students, which can lead to a fundamental understanding that is more likely to be retained.

A thorough presentation of structural mechanics theory and applications includes some of these topics:

        Properties of wood and lumber grades
        Tension member design
        Bending Theory
        Beam design
        Buckling Theory
        Compression member design
        Combined bending and axial force
        Engineered wood products
        Glulam beam and column design

CHAPTER

1

WOOD AS STRUCTURAL MATERIAL

1.1 Introduction

Wood has been used as a building material for thousands of years in its natural state. Wood construction is common for many single-family houses throughout the world.

In areas where timber and wood materials are easily accessible, wood construction is often considered to be the cheapest and best approach for small housing structures. In the USA, wood frame is used for approximately 90% of the houses constructed, predominantly in suburban regions. The ability to construct wood buildings with minimal use of specialized equipment has kept the cost of wood frame buildings competitive with other types of construction. Today, engineered wood is becoming very common in industrialized countries. Wood used as construction material should be carefully chosen because different types of wood have different qualities and structural properties. They should be used for the purposes for which they are best adapted. Wood is a durable, economical and beautiful construction material when it is properly chosen and used. In the USA, there are over 1,000 species of trees. Of these, only about 100 are used for constructing and manufacturing wood products.

1.2 Types of Wood

It is common to classify wood as either softwood or hardwood. The wood from conifers (e.g. pine) is called softwood, and the wood from deciduous trees (e.g. oak) is called hardwood. These names may be misleading, because hardwoods are not necessarily hard, and softwoods

are not necessarily soft. For example, balsa is a hardwood but is actually softer than any commercial softwood. Conversely, some softwoods are harder than many hardwoods. The tree's seeds determine whether the wood is considered hardwood or softwood. Hardwood trees have seeds with some sort of covering. For example, apple trees are hardwoods, since the apples cover the apple seeds. In contrast, softwood trees always have uncovered seeds that simply fall from the trees. For example, pine trees have cones that release their seeds directly into the air.

Softwoods

Softwoods or conifers, from the Latin word meaning "cone-bearing," have needles rather than leaves and are generally evergreen. Widely available U.S. softwood trees include cedar, fir, hemlock, pine, redwood and spruce. Most structural lumber is manufactured from softwoods because of the trees' faster growth rate, availability, and workability (i.e., ease of cutting and nailing).

Hardwoods

Hardwoods are deciduous trees that have broad leaves, produce a fruit or nut and generally go dormant in the winter. North America's forests grow hundreds of varieties that thrive in temperate climates, including oak, ash, cherry, maple and poplar species. Each species can be crafted into durable, long-lasting furniture, cabinetry, flooring and millwork, and each offers unique markings with variation in grain pattern, texture and color.

Key Points

	**Softwood** From Coniferous Trees (with needlelike leaves) Used for structural framing members Southern Pine, Douglas Fir
	**Hardwood** From Deciduous Trees (with broad leaves) Used for flooring and furniture Maple, Oak, Birch

1.3 **Characteristics of Wood**

Anisotropic Material

Wood is a biological material and is one of the oldest structural materials. It is composed of *cells*, whose walls are made up of **cellulose** and bonded together by a glue-like substance called lignin. Wood can be thought of as a cellular composite material. The cellular structure of wood gives relatively high strength-to-weight ratio of 2:1, comparative to 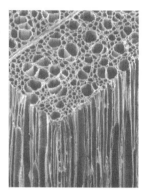 steel. Also, the cellular structure of wood and the physical organization of the cellulose chain within the cell wall make the physical and mechanical properties of wood highly dependent upon the direction of loading.

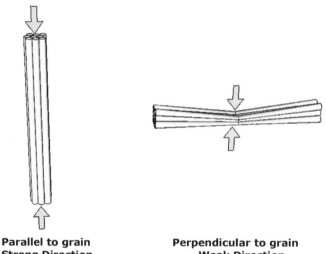

| Parallel to grain | Perpendicular to grain |
| Strong Direction | Weak Direction |

Figure above shows a simplified depiction of the anisotropic properties of wood, comparing it to a bundle of thin-walled drinking straws. Parallel to their longitudinal axes, the straws can support loads substantially greater than their weight. When loaded perpendicular to the longitudinal axis, the straws yield under much lower loads.

This means that it has an internal structure that is like a bundle of drinking straws which are actually hollow vertical cells and are effectively water carrying tubes. These vertical cells are held together normal to their length, creating a strong internal structure that allows efficient flexure through shearing action of the bundled cells. When loads are applied perpendicular to this internal structure, the loads can crush the hollow tubes relatively easily, thus, reducing the allowable stresses perpendicular to the grain. In other words, wood is orthotropic in its internal structure - different strength properties in three different orthogonal directions: longitudinal (strongest), tangential (weaker), and radial (weakest). To use wood to its best advantage and most effectively in engineering applications, its specific characteristics and physical properties must be taken into consideration.

1.4 Advantages & Disadvantages of Wood

Building materials are selected based on a number of factors in addition to durability, including cost, availability, ease of construction, thermal performance, and aesthetics. Wood performs equally or better compared with other building materials in many, if not all, of these categories. As with all materials, wood is susceptible to deterioration under specific adverse conditions. However, with proper design detailing, good construction techniques and adequate building maintenance, wood structures can be expected to last an exceedingly long time.

Advantages

High Strength-to-Weight Ratio

The specific strength is a material's strength divided by its density. It is also known as the strength-to-weight ratio. Because wood has a relatively high strength to weight ratio, dead load is a smaller component of the total load factor than it is for heavier materials. It is usually preferable to use the lightest or least involved construction type appropriate for a given span that is capable of carrying the design load.

Aesthetics

Nothing compares to word for the aesthetic "warmth". Wood has aesthetically pleasing characteristics that make it desirable in the construction of paneling, doors, cabinets and furniture.

Economics

Wood is the most affordable building material. The economy of wood construction is one of the many reasons why wood-frame construction has remained the preferred method for residential construction.

Easy Construction

Also, wood is easy to work with and lightweight, requiring only moderate construction skills. Its practicality and workability make construction simple and efficient for use in residential or small commercial applications.

Thermal Performance

In comparison with metal or concrete, wood conducts very little heat because heat moves very slowly between wood cells. For example, wood is 400 times less heat-conductive than steel.

Flexibility

Another quality that makes wood useful as a construction material is its flexibility. Buildings constructed using wood structural materials are capable of some movement under stress.

Disadvantages

Some qualities that detract from wood as a material for construction include its lack of uniformity. Variability even exists in lumber from the same tree. Wood also breaks down over time, especially when damp conditions prevail.

Variability in Structural Properties

Variation in properties, is common to all construction materials. However, because wood is a natural material and the tree is subject to many constantly changing influences (such as moisture, soil conditions, and growing space), structural properties of wood vary considerably, even in clear material.

Dimensional Instability - Shrinkage/Swelling

Environmental humidity, temperature, and age are all factors that can make wood change its volumn and shape.

Combustibility

Wood, of course, can burn-- making it a less than ideal material to use in applications where fire safety is a concern.

Decay or Wood-Destroying Pests

One of the greatest disadvantages of wood is that a variety of pests can destroy wood. These pests are difficult to identify, expensive to eradicate, and can do great amounts of damage that are virtually invisible until it's too late to repair.

Durability

Some types of wood coupled with different types of wood treatments have greatly enhanced the general durability of wood, but there are many applications for which wood is unsuitable due to durability issues.

Moisture

Moist conditions can, over time, soften even wood that has been treated to withstand moisture, making it susceptible to wet rot and fungus.

Summary

Advantage
1. High Strength-to-Weight Ratio
2. Aesthetics
3. Economical
4. Easy Construction
5. Thermal Performance

Disadvantage
1. Varying Structural Properties
2. Dimensional Instability - Shrinkage
3. Combustible - Fire
4. Decay or Woodworm Damages
5. Discoloration

1.5 **Moisture Content**

In addition to the solid material, wood also contains moisture. The amount of water contained in a wood member affects many structural properties of it. Moisture content is the weight of water as a percentage of the oven-dry weight of the wood.

$$Moisture\ Content = \frac{moist\ weight - oven\ dry\ weight}{oven\ dry\ weight} \times 100\ \%$$

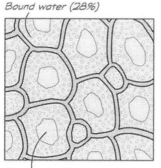

Bound water (28%)

Free water (72%)

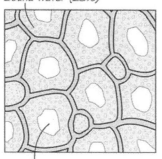

Bound water (28%)

Free water dissapated
Fiber Saturation Point

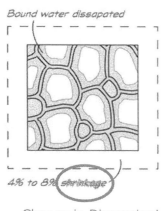

Bound water dissapated

4% to 8% shrinkage

Change in Dimension!

The water within the wall of cells is called *bound water* (approximately 28%), and the water in the cell cavity *free water* (approximately 72%). As the wood cell is exposed to relatively drier air, the free water leaves the cell easily, but the cell wall stays saturated. As long as the cell walls remain saturated, the piece of wood will not change in shape—it will not shrink or swell. But when the wood loses all of its free water, the bound water in the cell starts to evaporate. This is a critical point, called the *fiber saturation point* (**FSP**), when all of the free water has evaporated, but all of the bound water is still present. As the water continues to evaporate from the cell wall, the piece of wood begins to change in dimension—it shrinks. Wood will continue to lose water from the cell walls until it reaches equilibrium with the surrounding air. Wood scientists call this point *equilibrium moisture content* (**EMC**). Remember that the piece of wood continues to shrink as the bound water is removed below FSP. The EMC of wood is determined by the relative humidity of the surrounding air. The lower the relative humidity (the drier the air), the more water is evaporated from the piece of wood. If the surrounding air is extremely dry, as in a hot oven, the wood will eventually lose all of its water, or become "oven dry." Naturally, as the wood picks up water, it will swell and change shape until the FSP is reached. At this point, more water may be picked up, but no additional swelling will occur.

Seasoning

Seasoning is a controlled drying process of lumber to improve its structural properties. There are two common methods of drying lumber: air drying and kiln drying. Lumber can be purchased as green lumber (no significant drying has occurred since the lumber was cut from logs) or as air-dried or kiln dried lumber. All lumber increases in strength, hardness, and stiffness as it dries from green moisture content to lower moisture content.

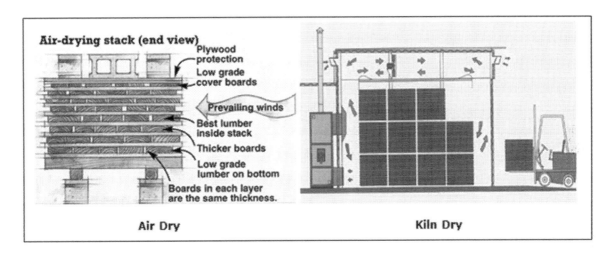

Air Dry   Kiln Dry

Moisture Designation

Most lumber is dried to the *S-DRY* (surfaced dry) condition, meaning that that the moisture content is less than 19%. A grade of S-GRN indicates that the lumber is surfaced green, and contains a moisture content of above 19%. Finally, MC 15 tells you that the wood contains less than 15% moisture content. These grades indicate moisture content when the wood is milled. Subsequent humidity or exposure to water can change the moisture content.

Moisture Designation	MC at time of Manufacture
S-GRN (Surfaced Green)	MC > 19%
S-DRY (Surfaced Dry)	MC ≤ 19%
MC 15	MC ≤ 15%

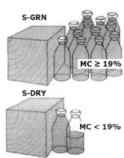

1.6 Effects of Moisture Content

Moisture can have a number of detrimental effects on the structural performance of timber such as:

Strength - Water in the cell walls acts as a lubricant and allows the fibers to slide past each other more easily. There is a small reduction in strength of wood fibers as the MC increases.

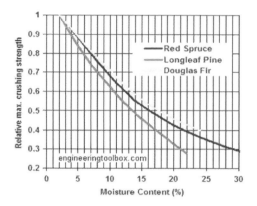

Stiffness - Water lubrication within the cells causes a small increase in elastic deflection under load (this is a decrease in stiffness). However, moisture has a marked effect on creep. With only loose bonds between the cells, as load is applied, the fibers rely on friction to stop them from sliding over each other. Under long term loading, some sliding will occur. This is referred to as creep.

Change in dimensions – The shrinking and swelling in dimension is not the same in all directions. Tangential (along growth rings) shrinkage is always a little larger than the shrinkage in the radial direction.

Durability - fungi and termites need moisture to thrive. Moist wood is therefore more vulnerable to biological degradation.

Coatings - unless the protective coatings are flexible, the shrinkage and swelling of timber as moisture moves in and out causes deterioration of the coatings. Once a coating has been broken, water can move into the timber.

1.7 Lumber Defects

Man Made Defects

Bow
A curve along the face of a board that usually runs from end to end.

Check
A crack in the wood structure of a piece, usually running lengthwise. Checks are usually restricted to the end of a board and do not penetrate as far as the opposite side of a piece of sawn timber.

Crook
Warping along the edge from one end to the other. This is most common in wood that was cut from the center of the tree near the pith.

Cup
Warping along the face of a board across the width of the board. This defect is most common of plain-sawn lumber.

Shake
Separation of grain between the growth rings, often extending along the board's face and sometimes below its surface.

Split
A longitudinal separation of the fibers which extends to the opposite face of a piece of sawn timber.

Twist
Warping in lumber where the ends twist in opposite directions.

Wane
The presence of bark or absence of wood on corners or along the length of a piece of lumber.

Natural Defects

Loose Knot

A knot that cannot be relied upon to remain in place in the piece. Caused by a dead branch that was not fully integrated into the tree before it was cut down.

Tight Knot

A knot fixed by growth or position in the wood structure so that it firmly retains its place in the surrounding wood.

Pitch

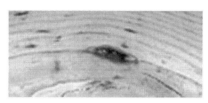

An accumulation of resinous material on the surface or in pockets below the surface of wood. Also called gum or sap.

Wormhole

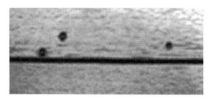

Small holes in the wood caused by insects such as beetles.

Blue Stain

A discoloration that penetrates the wood fiber. It can be any color other than the natural color of the piece in which it is found.

Spalt

Typically found in dead trees, spalting is any form of wood discoloration caused by fungi.

CHAPTER

2

SAWN LUMBER

2.1 Introduction

Sawn lumber is obtained by sawing round logs into rectangular shapes. All lumber directly sawn from a log is rough lumber and has surface imperfections caused by the initial sawing operation. The rough sawn size is usually called the nominal size. After seasoning, rough lumber is smoothed on sides and edges by a surfacing machine to become dressed lumber. Making lumber smooth also reduces the size (width and thickness). This new finished size is called the actual size. Nominal sizes are used by the construction industries and sawn lumber is classified by nominal size.

Classification of Sawn Lumber

Based on the cross-sectional dimensions, sawn lumber is classified into 3 different categories: dimension lumber, boards, and timbers. Dimension lumber is further subdivided into five categories based on size classifications. These classes are structural joists and planks, studs, decking, light framing, and structural light framing. Timbers are also subdivided into two groups by size classification: Beams and Stringers and Posts and Timbers. It must be noted that the lumber size classifications are based on the most efficient *anticipated* use of the member, rather than the actual use. But there are no restrictions on actual usage for any size classifications.

Size Classification of Lumber

1. **Dimension Lumber** (lumber with nominal thickness 2 to 4)
   - from 2×2 thru 4×16
   - Sectional properties are calculated based on the ***dry*** size

2. Boards
   - 1 to 1.5 inches thick, 2 inches and wider
   - MSR (Machine Stress-Rated) Board

3. **Timber** (lumber with nominal thickness 5″ or more)
   - a. **Beams and Stringers** (rectangles) – 5″ and thicker, width > 2″ + thickness
   - b. **Posts and Timbers** (squares) – 5x5 and larger, width ≤ 2″ + thickness
   - Sectional properties are based on the ***green*** size

2.2 **Wood Design Standards**

In the USA, the basic design criteria from which most wood structures are constructed, are specified in the 2005 National Design Specification for Wood Construction (**NDS**). All or part of the **NDS** is usually incorporated into the International Building Codes (**IBC**).

Originally created in the late 1940's, the **NDS** is the most nationally recognized design guide for wood structures. The foundation of this design guide is Allowable Stress Design (ASD), where stresses generated from design loads are compared to allowable material stresses. In general the **NDS** provides the requirements, design provisions and formulas for structural lumber, glued laminated timber, structural-use panels, shear walls and diaphragms, poles and piles, I-joists, structural composite lumber, and structural connections. In 2005, the NDS adopted a dual format approach which enables the designer to use either an allowable stress design (**ASD**) approach or a load and resistance design (**LRFD**) approach. The LRFD addresses code criteria based on a reliability approach. This approach compares the internal forces resulting from applied *factored loads* to the *resistance* of the specific member, connection, or assembly. Due to a code calibration process when the **LFRD** was developed, typically the **ASD** and **LRFD** result in the same member design. Currently, **ASD** is more popular choice among design professionals for wood design. For this reason, the **ASD** approaches are used in this textbook and **LRFD** version will be prepared and added to the textbook in near future.

Allowable Stress Design Method (ASD)

ASD used to be referred to as *Elastic Design* or *Working Stress Design* because ASD is based on elastic stresses of structural members subjected to working loads. In ASD approaches, the maximum stress in a structural member is calculated under service load conditions and must not exceed a code-specified allowable stress which includes a factor of safety. The design criteria for ASD can be expressed as:

$$\boxed{\textbf{Max. Actual Stress} \leq \textbf{Allowable Stress}}$$

Load and Resistance Factor Design (LRFD)

The Load Factors are multiplied to the service loads to provide more realistic safety margins. The Resistance of a member is determined based on limit states (fracture, yielding, buckling, etc.) to address uncertainties involved in design, manufacturing and construction processes. The theoretical resistance (or strength) of the member is reduced by the Resistance Factors or Strength Reduction Factors. The design criterion for LRFD is:

Σ (Load Factor × Load effect) ≤ Resistance Factor × Resistance

The load factors are usually greater than unity while the resistance factors are generally less than unity.

2.3 Reference Design Values

The design values tabulated in the NDS Supplement are referred to as *Reference Design Values*, or *Base Design Values*, and factors of safety have been reflected in these values, except modulus of elasticity. The symbol of an upper case F, is used for the Reference Design Values and a subscript is added to indicate the stress type. For example, F_b represents the reference design value for bending stress. Similarly, a symbol E, is given to the modulus of elasticity. In determining the allowable stress for a specific design, start with the Reference Design Values and multiply the applicable *adjustment factors*. An upper case with a prime is given to the allowable stress to indicate that the reference design value has been modified to account for different use conditions of the wood member. For example, F_b' represents the allowable bending stress. Below is a list of the stress types which are typically needed to design structural wood members. The Reference Design Values to be used depend on the types of stresses that the structural member experiences.

1. F_b (Bending Stress)

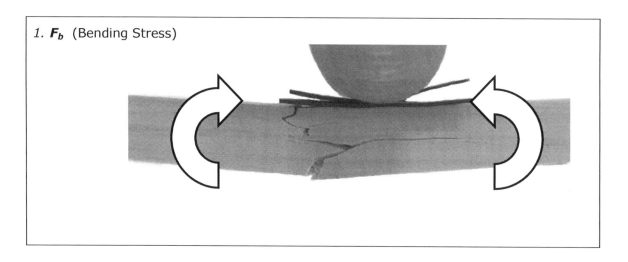

2. **F_c** (Compressive Stress parallel to grain)

Maximum stress sustained by a compression parallel-to-grain specimen having a ratio of length to least dimension of less than 11. In this way, the effects of buckling can be excluded.

3. **$F_{c\perp}$** (Compressive Stress perpendicular to grain)

$F_{c\perp}$ is generally less than F_c. Thus when two wood members meet with an angle, special consideration must be given. Base plates must be provided to avoid failures perpendicular to grain.

4. **F_v** (Shear Stress parallel to Grain)

Shear strength to resist internal slipping of one part upon another along the grain is the **most critical** in wood member design. Design values represent average strength in radial and tangential shear planes.

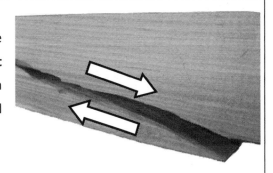

5. **F_c** (Tensile Stress parallel to grain)

Maximum stress sustained by a tensile force parallel-to-grain specimen having a ratio of length to least dimension of less than 11.

2.4 Adjustment Factors

Strength and performance of wood members can be affected by many factors such as moisture, duration of sustained loads, temperature, etc. as will be described in more detail below. To account for the service conditions that will degrade the structural properties of wood, the allowable stresses are determined by multiplying the *Reference Design Values* by the appropriate *adjustment factors*. Below is a list of the significant adjustment factors that are applied to the *Reference Design Values*. Not all of the adjustment factors are used on every reference design value. For example, the column stability factor, C_P and the buckling stiffness factor, C_T. Most adjustment factors will end up defaulting to 1.0 therefore it will not change the allowable design value. With practice, you will get used to which factors will need to be checked, and which can be ignored. The subscripts indicate the purpose of the adjustment and thus are case sensitive. An upper case must not be changed to a lower case, or vice versa.

Applicability of Adjustment Factors for Sawn Lumber   **Table A-1**

Adjusted Design Values	Base Values	ASD only	ASD and LRFD											LRFD only		
		Load duration factor	Size factor	Repetitive member factor	Temperature factor	Beam stability factor	Wet service factor	Flat use factor	Incising factor	Column stability factor	Buckling stiffness factor	Bearing area factor	Format conversion	Resistance factor	Time effect factor
$F_b' = F_b \times$		C_D	C_F	C_r	C_t	C_L	C_M	C_{fu}	C_i	–	–	–	K_F	ϕ_b	λ
$F_t' = F_t \times$		C_D	C_F	–	C_t	–	C_M	–	C_i	–	–	–	K_F	ϕ_t	λ
$F_v' = F_v \times$		C_D	–	–	C_t	–	C_M	–	C_i	–	–	–	K_F	ϕ_v	λ
$F_{c\perp}' = F_{c\perp} \times$		–	–	–	C_t	–	C_M	–	C_i	–	–	C_b	K_F	ϕ_c	λ
$F_c' = F_c \times$		C_D	C_F	–	C_t	–	C_M	–	C_i	C_P	–	–	K_F	ϕ_c	λ
$E' = E \times$		–	–	–	C_t	–	C_M	–	C_i	–	–	–	–	–	–
$E_{min}' = E_{min} \times$		–	–	–	C_t	–	C_M	–	C_i	–	C_T	–	K_F	ϕ_s	–

C_F – Size Factor

The size of sawn lumber has an effect on its strength because wood is not a homogeneous material. To take this non-homogeneity into account, the size factor (C_F) is used. In general, the larger a wooden member becomes, the lower the size factor will get. The size factor is used for sawn lumber design to calculate bending stress (F_b), tension stress (F_t), and compression stress (F_c).

Size Factors (C_F) for Dimension Lumber   **Table A-2**

Grades	Nominal Width	F_b		F_t	F_c	Other
		2" & 3" thick	4" thick			
Select Structural, No. 1 & Better, No. 1, No. 2, & No. 3	2", 3", & 4"	1.5	1.5	1.5	1.15	1.0
	5"	1.4	1.4	1.4	1.1	1.0
	6"	1.3	1.3	1.3	1.1	1.0
	8"	1.2	1.3	1.2	1.05	1.0
	10"	1.1	1.2	1.1	1.0	1.0
	12"	1.0	1.1	1.0	1.0	1.0
	14" & wider	0.9	1.0	0.9	0.9	1.0
Construction & Standard	2", 3", & 4"	1.0	1.0	1.0	1.0	1.0
Utility	2" & 3"	0.4	-	0.4	0.4	1.0
	4"	1.0	1.0	1.0	1.0	1.0
Stud	2", 3", & 4"	1.1	1.1	1.1	1.05	1.0
	5" & 6"	1.0	1.0	1.0	1.0	1.0
	8" & wider	Use No. 3 grade Reference Design Values and Size Factors				

Size Factors (C_F) for Timber   **Table A-3**

Nominal Depth	Net Depth	Size Factor $C_F = (12/d)^{1/9}$
12" or less	(varies)	1.0
14"	13.5"	0.987
16"	15.5"	0.972
20"	19.5"	0.947
22"	21.5"	0.937
24"	23.5"	0.928
26"	25.5"	0.920
28"	27.5"	0.912
30"	29.5"	0.905

Note : The size factor for timber is applied only to bending.

C_r - Repetitive Member Factor

The repetitive-member factor recognizes system performance similar to the idea that breaking one stick is easier than breaking a bundle of sticks. If a member is overloaded, the overload is distributed by sheathing to adjacent members. When multiple sawn lumber members act compositely to properly distribute a load amongst themselves, 15% increase in F_b is allowed to account for these beneficial system effects, or $C_r = 1.15$.

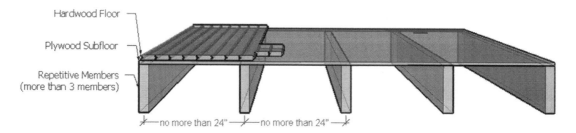

Conditions for **repetitive-member system**

1. Three or more parallel members of Dimension Lumber.
2. Members spaced no more than 24 inches on center.
3. Members connected together by a load-distributing element such as roof, floor or wall sheathing.

(If all of the above conditions are met, $C_r = 1.15$. Otherwise, $C_r = 1.0$)

C_D - Load Duration Factor

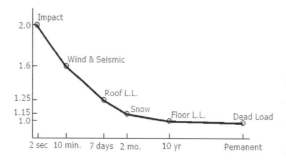

Wood can support higher stresses if the loads are applied for a short duration of time. In other words, wood has a higher strength when a load is applied instantaneously than it does when the load is applied for a long period of time. Duration of load is the total cumulative length of time that the full design load is applied. The NDS assumes that 10 years is the "normal" cumulative load duration for wood members and connections. Therefore, a factor of $C_D =$

1.0 applies for loads with a cumulative duration of ten years during the life of the structure. When load combinations include loads of shorter durations (e.g. - snow loads, wind loads, seismic loads), the C_D-value associated with the shortest-duration load in the given load combination is to be used.

Load Duration Factors   **Table A-4**

Load Type	C_D	Time Frame
Permanent (Dead Load)	0.9	Greater than 10 years
Normal (**Floor Live Load**)	**1.0**	**10 years**
Snow Load	1.15	2 months
Roof Live Load	1.25	7 days
Wind or Seismic Load	1.6	10 minutes
Impact Load	2.0	Less than 2 seconds

C_{fu} – **F**lat **U**se Factor

A wooden beam is used about its strong axis of the cross-section. When members are loaded in bending about the minor axis, the flat use factor may be used. In general, the flat use factor will increase the strength of a member, therefore it is conservative to take it as unity.

Major Axis Bending   **Minor Axis Bending**
                                         (Flat Use Factor to be used)

Flat Use Factors (C_{fu})   **Table A-5**

Nominal Width	Nominal Thickness	
	2" & 3"	4"
2" & 3"	1.00	N/A
4"	1.10	1.00
5"	1.10	1.05
6"	1.15	1.05
8"	1.15	1.05
10" & wider	1.20	1.10

C_L – Beam (**L**ateral) Stability Factor

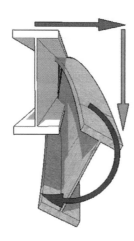

When the compression side of a beam is not laterally braced, the beam may buckle (lateral torsional buckling) at a lower bending stress level than the specified allowable stress. That means the bending moment strength of a beam is significantly reduced if the lateral torsional buckling is not prevented. Practical situations when the beam stability factor is not necessary, i.e., C_L = 1.0 is as follows:

✓ When the depth of the member does not exceed its width (d ≤ b),
✓ When rectangular sawn lumber is laterally braced in accordance with Section 4.4.1 of NDS,
✓ When the compression edge of a bending member is continuously laterally supported, and the ends have sufficient connections to prevent rotation,

Usually, a roof or floor diaphragm provides the compression side of beam with continuous lateral support. This will allow most floor and roof systems to be able to ignore the stability factor.

Otherwise, C_L is calculated as:

1. Determine the slenderness ratio, R_B :

$$R_B = \sqrt{\frac{l_e\, d}{b^2}}$$

Note: The slenderness ratio should not exceed 50.
where:
   l_e = Effective Unbraced Length (Found in NDS Table 3.3.3)
   d = depth of the member
   b = width of the member

2. Determine the critical buckling design value, F_{bE} :

$$F_{bE} = 1.2\frac{E'_{min}}{(R_B)^2}$$

where:

E'_{min} = E_{min} x C_M x C_t x C_i x C_T
E_{min} = Minimum modulus of elasticity
R_B = The Slenderness ratio determined in Part 1.

3. Solve for F_b*:

F_b* is a reference bending design value as follows:

For Sawn Lumber:

$$F_b* = F_b \times C_D \times C_M \times C_t \times C_F \times C_i \times C_r$$

For Glued Laminated Lumber:

$$F_b* = F_b \times C_D \times C_M \times C_t \times C_c$$

4. Solve for C_L:

$$C_L = \frac{1 + \frac{F_{bE}}{F_{bx}*}}{1.9} - \sqrt{\left(\frac{1 + \frac{F_{bE}}{F_{bx}*}}{1.9}\right)^2 - \frac{\frac{F_{bE}}{F_{bx}*}}{0.95}}$$

C_M – Wet Service Factor (**M**oisture Factor)

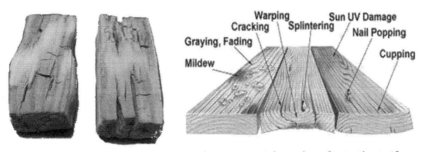

The Wet Service Factor is used to consider the fact that if a wooden member cycles between wet/dry conditions, it will degrade in condition faster than one that is subject to continuous wet or dry conditions. For example, a column supporting a wood pier under water will retain its strength longer than the part of the column that is stuck between the tides. This is because the transition between wet and dry will deteriorate the wood faster than a constantly wet, or constantly dry member. For sawn lumber with the EMC (equilibrium moisture content) exceeding 19%, $C_M <$ 1.0 as follows:

Wet Service Factors (C_M) for Sawn Lumber   Table A-6

use with	F_b [1]	F_t	F_c [2]	F_v	$F_{c\perp}$	E & E_{min}
C_M	0.85 [1]	1.0	0.8 [2]	0.97	0.67	0.9

[1] when $(F_b \times C_F) \leq 1150$ psi, $C_M = 1.0$
[2] when $(F_c \times C_F) \leq 750$ psi, $C_M = 1.0$

For glulam members with EMC exceeding 16%, C_M < 1.0 as follows:

Wet Service Factors (C_M) for Glulam						Table A-7
use with	F_b	F_t	F_c	F_v	$F_{c\perp}$	E & E_{min}
C_M	0.8	0.8	0.73	0.875	0.53	0.833

C_i – **I**ncising Factor

In order for sawn lumber to be used in outdoor conditions, a preservative treatment is required to prevent decay in the wood. This process includes permeating the wood with chemical preservatives to prevent decay-causing fungus from entering the

wood. In most cases, this is made possible by pressure treatment, but certain species of wood must be incised (small slits into the wood), to receive the chemicals more readily. The loss of the cross-sectional area and section modulus from the small incisions of a wood member will require certain reduction factors to be used.

C_f – **F**orm Factor

Round Section
$C_f = 1.18$

Rhombus Section
$C_f = 1.414$

The form factor, C_f has been in the specification for wood design for many years, but it has not been used often. The purpose of the form factor is to adjust the reference design value for bending stress, F_b for certain cross sections that are not rectangular shapes.

C_t – Temperature Factor

The strength of a member is decreased as the service temperature is higher than normal temperatures. However, reductions in strength caused by heating below 150° F are generally reversible. For wood that is expected to be exposed to high temperatures for long periods of time, a temperature factor will need to be taken into account when designed. For temperatures below 100° F, C_t is equal to unity (1.0). The following temperature factors can be used for sawn lumber, glued laminated timber, prefabricated I-joists, structural composite lumber, and wood structural panels.

Temperature Factors (C_t)   Table A-8

Values	Moisture Conditions	C_t		
		T ≤ (100°F)	(100°F ≤ T ≤ 125°F)	(125°F ≤ T ≤ 150°F)
F_t, E, E_{min}	Wet or Dry	1.0	0.9	0.9
F_b, F_v, F_c, $F_{c\perp}$	Dry	1.0	0.8	0.7
	Wet	1.0	0.7	0.5

C_b – Bearing Area Factor

The bearing area factor (C_b) is used to increase design values for concentrated loads on wood perpendicular to the grain. The factor applies to bearings of any length at the ends of the member, and to all bearings 6" or more in length at any other location.

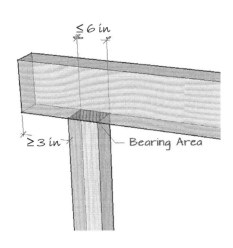

It is useful when high loads are present on washers, hangers, etc. Note: If all the previous conditions are not met, or if the bearing area is greater than 6", then the bearing area factor can be taken as unity (1.0). For bearing lengths, L_b, less than 6" and not nearer than 3" to the end of a member, $F_{c\perp}$ shall be permitted to be multiplied by following bearing area factor, C_b.

Bearing Area Factors (C_b)							Table A-9
L_b	0.5"	1"	1.5"	2"	3"	4"	6" or more
C_b	1.75	1.38	1.25	1.19	1.13	1.10	1.00

C_P – Column Stability Factor

The column stability factor is to ensure that weak-axis buckling or torsional buckling does not occur over unbraced length. Note that this is similar to the beam stability factor except it deals with columns in compression, vs. beams in flexure. When the column is supported throughout its length, C_P can be unity, 1.0.

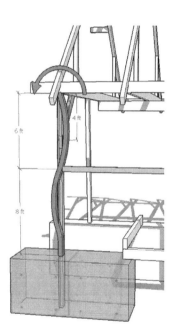

1. *Controlling slenderness ratio*

$$\left(\frac{Kl}{d}\right)_x \quad \text{or} \quad \left(\frac{Kl}{d}\right)_y$$

where:
  l = Unbraced length of the column
  K = Effective Length Factor

Note that d (depth of the member) is used in place of radius of gyration (r) for rectangular beams. If other column shapes are used, d in the procedure above should be replaced with $r\sqrt{12}$. Also note that the slenderness ratio should not exceed 50. The NDS allows a slenderness ratio of up to 75 during construction. But once the building is finished, and ready for occupancy, something must be done to keep the ratio down to or below 50. This is typically (and simply) done with the application of wall covering or sheathing. It may also be accomplished by bridging.

2. *Critical buckling design value,* **F_{cE}**

$$F_{cE} = 0.822 \frac{E'_{min}}{\left(\dfrac{Kl}{d}\right)^2}$$

Where,

E'_{min} = Minimum modulus of elasticity modified by adjustment factors

3. *Modified design value,* **F_c***

F_c* is a reference compression design value modified by all modification factors except **C_P**.

For Sawn Lumber:	**$F_c* = F_c \,(C_F)(C_D)(C_M)(C_R)(C_t)(C_i)$**
For Glued Laminated Lumber:	**$F_c* = F_c \,(C_F)(C_D)(C_M)(C_t)$**
For Round Timber Poles/Piles:	**$F_c* = F_c \,(C_D)(C_t)(C_u)(C_{cs})(C_{sp})$**

4. *Column Stability Factor,* **C_P**

$$C_P = \frac{1+\dfrac{F_{cE}}{F_c^*}}{2c} - \sqrt{\left(\frac{1+\dfrac{F_{cE}}{F_c^*}}{2c}\right)^2 - \frac{\dfrac{F_{cE}}{F_c^*}}{c}}$$

where,

c =	0.8	for sawn lumber
	0.85	for round timber pole
	0.9	for glulam timber

Note that it may be beneficial to calculate the value of $\dfrac{F_{cE}}{F_c^*}$ first. The **C_P**-values are tabulated in terms of $\dfrac{F_{cE}}{F_c^*}$ in Table B-5 in Appendix B.

C_V – Volume Factor

It has been noted that the reference design values in a wood member is affected by the relative size of the member. In sawn lumber, the size effect is taken into account by the size factor **C_F**. However, full-scale test data indicate that the size effect in glulam is related to the volume of the member rather than to only its depth. Therefore, the volume factor C_V replaces the size factor C_F for use in glulam design. Note that C_V is applied only to glulam beams subjected to loads applied perpendicular to the wide face of the lamination. The Reference Design Values of F_b are based on a 'standard size' glulam beam with following dimensions: width = 5 ⅛ in., depth = 12 in., length = 21 ft.

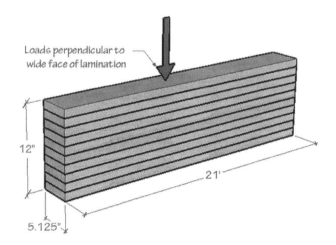

Standard Size of Glulam Beam

$$C_V = \left(\frac{5.125\ in}{b}\right)^{\frac{1}{x}} \left(\frac{12\ in}{d}\right)^{\frac{1}{x}} \left(\frac{21\ ft}{L}\right)^{\frac{1}{x}} \leq 1.0$$

where,

      b = Width of beam, in.
      d = Depth of beam, in.
      L = Length of beam between points of zero moment, ft.
      x = 20 for Glulam manufactured from Southern Pine
          = 10 for Glulam manufactured from all other species

C_T – Buckling Stiffness Factor

Buckling stiffness factor, C_T, Applies only to maximum 2x4 dimension lumber in the top chord of wood trusses that are subjected to combined flexure and axial compression.

2.5 Adjustment Factors for LRFD

The following three adjustment factors are applicable for only the LRFD approach.

K_F – Format Conversion Factor

The format conversion factor (K_F) is used to adjust reference ASD design values to LRFD reference resistances. By using LRFD in lieu of ASD design you are substituting three variables for one. Therefore, I normally recommend ASD as being the superior choice for wood design.

Format Conversion Factors (K_F)   **Table A-10**

Application	Property	K_F
Members	F_b	2.54
	F_t	2.70
	F_v F_{rt} F_s	2.88
	F_c	2.40
	$F_{c\perp}$	1.67
	E_{min}	1.76
Connections	(all connections)	3.32

Note: ϕ = resistance factor. K_F is not applicable where the LRFD design values are determined in accordance with the reliability normalization factor method (ASTM D 5457).

ϕ– Resistance Factor

The resistance factor (ϕ), also referred to as the *strength reduction factor*, is used to account for all uncertainties related to the material imperfection, manufacturing, construction and structural design procedures that may cause the strength to be lower than the theoretical values. The resistance factor is a function of the mode of failure.

Resistance Factors (ø)   **Table A-11**

Application	Property	Symbol	Value
Members :	F_b	ϕ_b	0.85
	F_t	ϕ_t	0.80
	F_v, F_{rt}, F_s	ϕ_v	0.75
	$F_c, F_{c\perp}$	ϕ_c	0.90
	E_{min}	ϕ_s	0.85
Connections:	(all types)	ϕ_z	0.65

λ– Time Effect Factor

The time effect factor (λ) is analogous to the **load** duration factor in principle (which is for ASD only), except this factor instead of being based on time duration (e.g. 7 days, 1 yr., etc.) is based on what load combination is being applied (which one should be able to associate the various load combinations with a time duration if they were so inclined).

Time Effect Factors (λ)   **Table A-12**

Load Combination	λ
1.4(D+F)	0.6
1.2(D+F) + 1.6(H) + 0.5(L_r or S or R)	0.6
1.2(D+F) + 1.6(L+H) + 0.5(L_r or S or R)	0.7 (L is from storage)
	0.8 (L is from occupancy)
	1.25 (L is from impact)
1.2D + 1.6(L_r or S or R) + (L or 0.8W)	0.8
1.2D + 1.6W + L + 0.5(L_r or S or R)	1.0
1.2D + 1.0E + L + 0.2S	1.0
0.9D + 1.6W + 1.6H	1.0
0.9D + 1.0E + 1.6H	1.0

Notes:
1. Time effect factors exceeding 1.0 are not applicable to: connections, pressure borne structural members, or structural members treated with fire retardant chemicals.
2. Load combinations shown above are in accordance with ASCE 7-02. Dependent on your own design codes the above load combinations may change.

Interpretation of Table A-12 for simplicity:

Table A-13 is not based on the NDS but rather author's interpretations of the above information.

Time Effect Factors based on Duration   **Table A-13**

Time Frame	λ	Example
Greater than 10 years	0.6	Permanent
10 years	0.7	Normal (Floor LL)
2 months	0.8	Snow Load
10 minutes	1.0	Wind or Seismic Load
Less than 2 seconds	1.25	Impact Load

2.6 Grades of Sawn Lumber

Lumber has quite variable structural properties because wood is a biological material. Wood members of similar structural properties are grouped into various stress grades such as Select Structural, No.1 and Better, No. 1, No, 2, Utility etc. Lumber is graded in accordance with standardized grading rules that consider the effect of natural growth characteristics and defects, such as knots and angle of grain, on the member's structural properties. Most lumber is visually graded, although it can also be machine stress-rated or machine evaluated.

Visually Grading

Majority of lumber is graded and sorted by visual inspection. Visual grading is performed by an experienced and certified grader in accordance with the grading rules of an approved grading agency. Stress grading determines strength and structural capacity. The grade of a wood member decreases as the number of growth defects increases and as their locations become more critical. For example, edge knots are more detrimental than knots located close to the neutral axis.

Grade	Grade Characteristics
Dense Select Structural Select Structural Select Structural NonDense	High quality, limited in characteristics that affect strength or stiffness. Recommended for uses where high strength, stiffness and good appearance are desired.
No.1 Dense No.1 No. 1 NonDense	Recommended for construction where high strength, stiffness and good appearance are desired.
No.2 Dense No.2 No.2 NonDense	Recommended for most general construction uses where moderately high design values are required. Allows well-spaced knots of any quality.
No.3	Recommended for general construction purposes where appearance is not a controlling factor. Many pieces included in this grade would qualify as No.2 except for a single limiting characteristic.

Typical Grade Stamps for Visually Graded Lumber

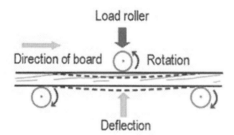

Machine Stress Rating (MSR)

A relatively small percentage of lumber is graded automatically by machine as well as a visual check. In this rating process, a bending test is performed to determine Modulus of Elasticity.

Load roller

Direction of board   Rotation

Deflection

Machine stress rating is based on a close co-relationship between the bending strength and the modulus of elasticity of lumber. To avoid permanent deformation and resulting residual stresses from the non-destructive bending test, MSR is limited to thin lumber (2 inches or less in thickness). Because the variability of MSR lumber is lower than visually graded lumber, it is often used for engineered wood products, e.g., glulam beams, light frame trusses and wood I joists, etc.

Grade Stamps

The grade stamp used by a grading agency helps the engineer, architect and contract to assure that the quality of lumber conforms to the contract specifications for the project.

Typical Grade Stamps for Machine Stress-Rated Lumber

MSR grades are typically specified in terms of their bending stress and stiffness. For example, the above stamp indicates the bending stress (F_b) of 1650 psi and the modulus of elasticity (E) of 1.5×10^6 grade.

Species and Species Groups

From a practical standpoint, different individual species with similar structural properties are grouped into commercial species groups. Typical species groups of structural lumber are shown below.

Species Group	Abbreviations	Individual Species Included
Douglas Fir - Larch	**DF-L**	Douglas Fir Western Larch
Douglas Fir - South	**DF-S**	Douglas Fir-South
Hem-Fir	**H-F**	California Red Fir Grand Fir Noble Fir Pacific Silver Fir Western Hemlock White Fir
Hem - Fir - (North)	**Hem - Fir (N)**	Pacific - Hemlock Amablis Fir
Spruce - Pine - Fir	**S - P - F**	Spruce (except Coastal Sitka) Jack Pine Lodgepole Pine Balsam Fir Alpine Fir
Western Cedars	**W-C**	Incense Cedar Western Red Cedar Prot Orford Cedar Alaskan Cedar
Western Woods	**W-W**	Alpine Fir Ponderosa Pine Sugar Pine Idaho White Pine Mountain Hemlock

2.7 Workshop for **Adjusted Design Values**

Case Study 2.1a   Determine the allowable design stresses for the 2x8 joists shown below.

Design Data :
1. Use No. 1 Douglas Fir–Larch.
2. Lateral bucking is not a concern.
3. Equilibrium moisture content (EMC) exceeds 19%.   - *Wet-Service Condition*
4. Normal temperature condition
5. The Controlling Load combination is **DL + RLL.**   - *Dead Load + Roof Live Load*
6. Unless otherwise specified, all other conditions are considered as "normal".

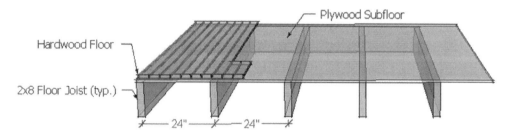

Solution

Reference Design Values – from **Table (B-1)**

	Grade	F_b	F_t	F_v	$F_{c\perp}$	F_c	E
D.F.-Larch	No. 1	1,000	675	180	625	1,500	1,700,000

Adjustment Factors

A. Size Factor (**C_F**) – from **Table (A-2)**

	for use with F_b	for use with F_t	for use with F_c
C_F for 2 × 8	1.2	1.2	1.05

B. Load Duration Factor (**C_D**) – from **Table (A-4)**

C_D for **DL**   = 0.9

C_D for **RLL** = 1.25 ← Largest one (the one for shortest duration) governs.

∴ **C_D** for the load combination **DL + RLL** = 1.25

(Note : Do not add the C_D's but pick the **largest** one!)

C. Repetitive-Member Factor (C_r)

Is this a repetitive-member system?

1. Three or more parallel members of Dimension Lumber. → Yes.

2. Members spaced not more than 24 inch on center. (∴ C_r is not for timbers or glulam.) → Yes.

3. Members connected together by a load-distributing element. → Yes.

Applicable : ∴ C_r = 1.15

D. Wet Service Factor (C_M) – from **Table** (A-6)

	for use w/ F_b^*	for use w/ F_t	for use w/ F_v	for use w/ $F_{c\perp}$	for use w/ F_c^{**}	for use w/ E
C_M	0.85*	1.0	0.97	0.67	0.8**	0.9

* $(F_b)(C_F)$ = (1000)(1.2) = 1200 > **1150** Yes? then use **0.85** (otherwise, C_M = 1.0)

** $(F_c)(C_F)$ = (1500)(1.05) = 1575 > **750**   Yes? then use **0.8**   (otherwise, C_M = 1.0)

E. Fire Retardant Factor (C_R)   Not applicable : ∴ C_R = 1.0

F. Temperature Factor (C_t)   Normal Temperature Condition   ∴ C_t = 1.0

G. Flat Use Factor (C_{fu})   Not applicable : ∴ C_{fu} = 1.0

H. Compression perpendicular to grain ($C_{c\perp}$)   0.04" deformation basis :   ∴ $C_{c\perp}$ = 1.0

I. Incising Factor (C_i)   No incision for preservative treatment assumed :   ∴ C_i = 1.0

J. Stability Factor (C_L)   No lateral buckling - Lateral supports provided by floor deck : ∴ C_i = 1.0

Allowable Stress Calculation

	Adjustment Factors										Allowable Stresses
	C_F	C_D	C_r	C_M	C_R	C_t	C_{fu}	$C_{c\perp}$	C_i	C_L	
F_b = 1,000	1.2	1.25	1.15	0.85	1.0	1.0	1.0	X	1.0	1.0	F'_b =1466 psi
F_t = 675	1.2	1.25	X	1.0	1.0	1.0	X	X	1.0		F'_t =1013 psi
F_v = 180	X	1.25	X	0.97	1.0	1.0	X	X	X	X	F'_v =218 psi
$F_{c\perp}$ = 625	X	X	X	0.67	1.0	1.0	X	1.0	X	X	$F'_{c\perp}$=419 psi
F_c = 1,500	1.05	1.25	X	0.8	1.0	1.0	X	X	1.0	X	F'_c =1575 psi
E = 1,700,000	X	X	X	0.9	1.0	1.0	X	X	1.0	X	E' =1530000 psi

SAWN LUMBER

Case Study 2-1b   Determine the allowable design stresses for the 2x8 joists shown below.

Design Data :
1.   Use No. 2 Douglas Fir–Larch South.
2.   Lateral bracing is provided for the compression sides of the joists.
3.   Equilibrium moisture content (EMC) exceeds 19%.   - Wet-*Service Condition*
4.   Normal temperature condition
5.   The Controlling Load combination is **DL + SL.**   - *Dead Load + Snow Load*
6.   Unless otherwise specified, all other conditions are considered as "normal".

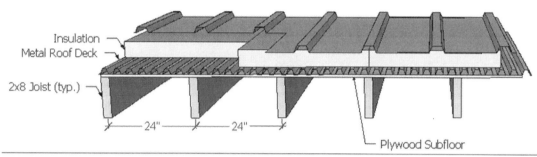

Solution

Reference Design Values – from **Table** (   )

Species	Grade	F_b	F_t	F_v	$F_{c\perp}$	F_c	E

Adjustment Factors

   A.   Size Factor (**C_F**) – from **Table** (   )

	for use with F_b	for use with F_t	for use with F_c
C_F for ( )			

   B.   Load Duration Factor (**C_D**) – from **Table** (   )

   C_D for **DL**   =

   C_D for **RLL** =

   ∴ **C_D** for the load combination **DL + RLL** =

(Note : Do not add the C_D's but pick the **largest** one!)

A. Repetitive-Member Factor (C_r)

    Is this a repetitive-member system?

      1. Three or more parallel members of Dimension Lumber. → Yes / No.

      2. Members spaced not more than 24 inch on center.   → Yes / No

      3. Members connected together by a load-distributing element. → Yes / No

          Applicable / Non-Applicable :   ∴ C_r =

B. Wet Service Factor (C_M) – from **Table** (   )

	for use w/ F_b*	for use w/ F_t	for use w/ F_v	for use w/ $F_{c\perp}$	for use w/ F_c**	for use w/ E
C_M						

* $(F_b)(C_F) = ($ $)($ $) = $ **> 1150** Yes? then $C_M = 0.85$ (otherwise, $C_M = 1.0$)

** $(F_c)(C_F) = ($ $)($ $) = $ **> 750** Yes? then $C_M = 0.8$ (otherwise, $C_M = 1.0$)

C. Fire Retardant Factor (C_R)

D. Temperature Factor (C_t)

E. Flat Use Factor (C_{fu})

F. Compression perpendicular to grain ($C_{c\perp}$)

G. Incising Factor (C_i)

H. Stability Factor (C_L)

Allowable Stress Calculation

Reference Design Values	Adjustment Factors										Allowable Stresses
	C_F	C_D	C_r	C_M	C_R	C_t	C_{fu}	$C_{c\perp}$	C_i	C_L	
F_b =								X			F'_b =
F_t =			X					X	X		F'_t =
F_v =	X		X				X	X	X	X	F'_v =
$F_{c\perp}$ =	X	X	X				X		X	X	$F'_{c\perp}$ =
F_c =			X				X	X		X	F'_c =
E =	X	X	X				X	X		X	E' =

Workshop 2-1a   Determine the allowable design stresses for the 4x10 joist member given below.

Design Data :
1. Use No. 2 Douglas Fir–Larch.
2. Lateral bucking is not a concern.
3. Equilibrium moisture content (EMC) exceeds 19%.   - Wet-*Service Condition*
4. Normal temperature condition
5. The Controlling Load combination is **DL + SL.**   - *Dead Load + Snow Load*
6. Shear Stress Factor is assumed to be 1.0.

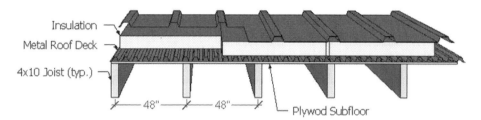

Insulation
Metal Roof Deck
4x10 Joist (typ.)
48"   48"
Plywod Subfloor

Solution

Allowable Stress Calculation

Reference Design Values	Adjustment Factors										Allowable Stresses
	C_F	C_D	C_r	C_M	C_R	C_t	C_{fu}	$C_{c\perp}$	C_i	C_L	
F_b =								×			F'_b =
F_t =			×				×	×			F'_t =
F_v =	×		×				×	×	×	×	F'_v =
$F_{c\perp}$ =	×	×	×				×		×	×	$F'_{c\perp}$ =
F_c =			×				×	×		×	F'_c =
E =	×	×	×				×	×		×	E' =

Workshop 2-1b   Determine the allowable design stresses for the 2x12 joist members given below.

Design Data :
1.   Use Select Structural Douglas Fir–Larch (South).
2.   Lateral bucking is not a concern.
3.   Equilibrium moisture content (EMC) ≤ 19%.   - Dry-*Service Condition*
4.   Normal temperature condition
5.   The Controlling Load combination is **DL + WL.**   - *Dead Load + Wind Load*

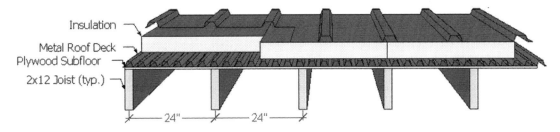

Insulation
Metal Roof Deck
Plywood Subfloor
2x12 Joist (typ.)
24" 24"

Solution

Reference Design Values	Adjustment Factors										Allowable Stresses
	C_F	C_D	C_r	C_M	C_R	C_t	C_{fu}	$C_{c\perp}$	C_i	C_L	
F_b =								×			F'_b =
F_t =			×				×	×			F'_t =
F_v =	×		×				×	×	×	×	F'_v =
$F_{c\perp}$ =	×	×	×					×		×	$F'_{c\perp}$ =
F_c =			×				×	×		×	F'_c =
E =	×	×	×				×	×		×	E' =

Workshop 2-1c   Determine the allowable design stresses for the member given below.

Design Data :
1. Use No. 1 Hem-Fir.
2. Lateral bucking is not a concern.
3. Equilibrium moisture content (EMC) doesn't exceeds 19%.
4. Normal temperature condition
5. The Controlling Load combination is **DL + EQ.**   - Dead Load + Seismic

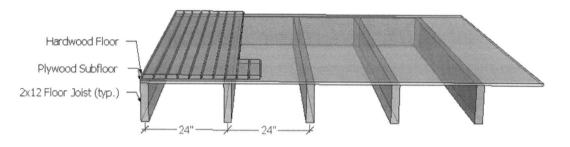

Hardwood Floor
Plywood Subfloor
2x12 Floor Joist (typ.)
24"   24"

Solution

Reference Design Values	Adjustment Factors										Allowable Stresses
	C_F	C_D	C_r	C_M	C_R	C_t	C_{fu}	$C_{c\perp}$	C_i	C_L	
F_b =								×			F'_b =
F_t =			×				×	×			F'_t =
F_v =	×		×				×	×	×	×	F'_v =
$F_{c\perp}$ =	×	×	×				×		×	×	$F'_{c\perp}$ =
F_c =			×				×	×		×	F'_c =
E =	×	×	×				×	×		×	E' =

Workshop 2-1d Determine the *allowable design bending stress (F_b')* for each member given below.
Note : When a use condition is not specified, consider it as 'normal use'.

1. Design Data :   a. Douglas Fir-Larch No. 1 Grade

                                     b. 2×8 roof joists spaced at 24 in. on center.

                                       c. Load combination to be considered is **DL + RLL**

	F_b	C_F	C_D	C_r	C_M	C_R	C_t	C_{fu}	C_i	C_L	
$F_b' =$	1000 psi	1.2	1.2	1.15	1.0	1.0	1.0	1.0	1.0	1.0	1725 psi

2. Design Data :   a. Douglas Fir-Larch No. 2 Grade

                                       b. 2×10 roof joists spaced at 24 in. on center.

                                       c. Load combination to be considered is **DL + SL**

	F_b	C_F	C_D	C_r	C_M	C_R	C_t	C_{fu}	C_i	C_L	
$F_b' =$											

3. Design Data :   a. Hem-Fir No. 1 Grade

                                       b. 6×16 floor beams.

                                       c. Load combination to be considered is **DL + LL**

	F_b	C_F	C_D	C_r	C_M	C_R	C_t	C_{fu}	C_i	C_L	
$F_b' =$											

4. Design Data :   a. Douglas Fir-Larch No. 1 Grade

                                  b. 6×8 columns.

                                    c. Load combination to be considered is **DL + LL + WL**

	F_b	C_F	C_D	C_r	C_M	C_R	C_t	C_{fu}	C_i	C_L
$F_b' =$										

5. Design Data :   a. Hem Fir No. 2 Grade

                                    b. 4×10 roof beams spaced at 48 in. on center.

                                    c. Load combination to be considered is **DL + SL**

	F_b	C_F	C_D	C_r	C_M	C_R	C_t	C_{fu}	C_i	C_L
$F_b' =$										

5. Design Data :   a. Douglas-Fir South No. 2 Grade

                                    b. 6×12 roof beams spaced at 6 ft. on center.

                                    c. Load combination to be considered is **DL + SL**

	F_b	C_F	C_D	C_r	C_M	C_R	C_t	C_{fu}	C_i	C_L
$F_b' =$										

TENSION MEMBER DESIGN

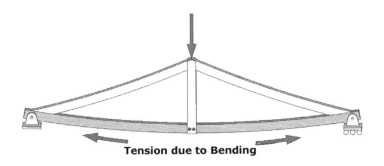

Tension due to Bending

3.1 Introduction

Axial tension members are found in many structural applications. For a typical example, the bottom chord of a truss subjected to gravity loads is generally in tension while the top chord is in compression. However, it should be noted that bending stress must be taken into consideration when the loads are not applied at joint locations or the joints can't be modeled as pins. In the design of tension members, any reduction of cross-section of a member (for example, bolt holes and notches) must be addressed because it creates the 'weakest link' of the tension member. The *net cross-sectional area* must be obtained by deducting the projected areas of holes or grooves from the *gross area*. Customarily, the projected area for nails is neglected in calculating the net area of a tension member. It must be emphasized that the tensile strength perpendicular-to-grain of wooden members is very low and stressing wood in tension across the grain must be avoided whenever possible. For this reason, our discussion in this chapter will be limited to the wooden members subjected to tension parallel-to-grain.

3.2 Net Area at Connections

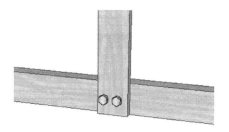 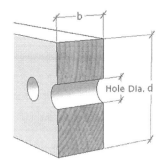

Bolted Connection   **Net Cross-sectional Area**

If a tension member is connected with bolt fasteners, its gross section is reduced because of the bolt holes. The critical section is in the section that has the minimum net area because the stress in a tension member is uniform throughout the cross-section except near the point of application of load, and at the cross-section with holes for bolts or other discontinuities, etc. So, the tensile stress of a tension member is calculated using the following formula:

$$f_t = \frac{P}{A_n}$$

where, **P** = Tensile force

A_n = Net cross-sectional area

= $b\,(d - D_h)$

In determination of net area, the diameter of a bolt hole should be taken 1/16 inches larger than the bolt diameter. Although the gross section for a tension member without holes should be taken normal to the direction of applied stress, the net section for a tension member with holes should be chosen as the one with the smallest area that passes through any chain of holes across the width of the member. Net section for a member with a chain of holes extending along a diagonal or zigzag line is the product of the net width (**d**) and thickness (**b**). To determine net width, deduct from the gross width the sum of the diameters of all the holes in the chain, then add, for each gage space in the chain.

3.3 **Load Path**

Every load applied to a building will travel (or flow) through the structural system until it is transferred to the supporting soil. The path that a load travels through is called a load path (or load flow). Of course, each structural element in the structural system must be designed for all loads that pass through it. All buildings need a proper load path to get the forces to the ground. If you don't provide it, Mother Nature and gravity will do it for you. The structural design is performed by chasing all loads throughout their load paths.

Load Path **shown by directional arrows on an actual structure**

3.4 **Tributary Area**

A tributary area of a structural element (such as a joist, beam, column, or wall) is the area of a loaded surface that contributes load to a particular member. The tributary area usually represents half the area all around the supporting element to the next supporting element. In the example of a truss, the actual loads in reality are usually imposed uniformly to the top chords and/or bottom chords of it. However, the real uniform loads are modeled as a series of concentrated loads applied only at the joint locations in a truss analysis. To obtain the joint loads, the concept of *tributary area* is introduced. The tributary area in this case refers to the roof area which transfers its loads to the particular joint of the truss.

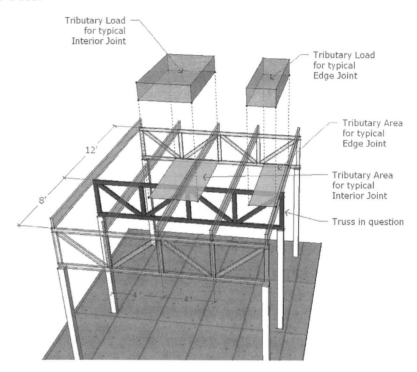

For the above truss, the tributary width for the middle truss will be the sum of halves the distances to the adjacent trusses, which is 4 ft + 6 ft = 10 ft.

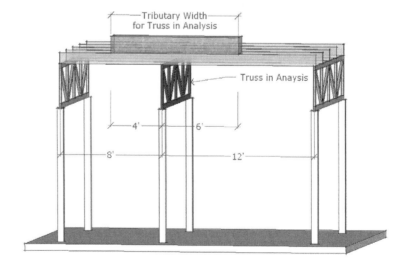

So, the actual uniform area load, 50 lb/ft^2 is multiplied by the tributary width of the truss, 10 ft to be modeled as a line load of 500 lb/ft or 0.5 k/ft.

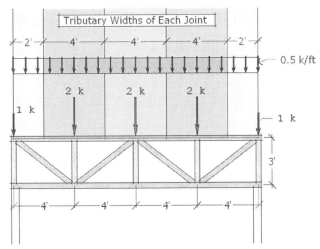

In turn, the uniform line load of 0.5 k/ft is multiplied by the tributary width of a typical internal joint, 4 ft, to get the point load of 2 k. In summary, a typical middle joint of the truss has a tributary area of 10 ft x 4 ft = 40 ft^2 which is subjected to the uniform area load of 50 lb/ft^2 and thus, its joint load will be 40 ft^2 x 50 lb/ft^2 = 2000 lbs = 2 kips. Similarly, the tributary area for each column can be visualized by color as shown below.

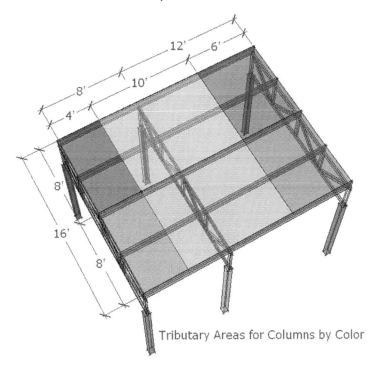

Tributary Areas for Columns by Color

3.5 Workshop for **Tension Member Analysis and Design**

Case Study 1a Determine the adequacy of the No. 1 Spruce Pine Fir, 2 x6 bottom (tension) chord in the truss shown.

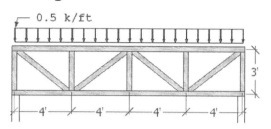

1. Load combination, DL + SL, controls.
2. No ceiling attached to the bottom chord
3. Connections : Single row of 3/4-in.-diameter bolts
4. The uniform load transferred from the roof system to the top chords is 0.5 k/ft.

Line loads to Joint loads
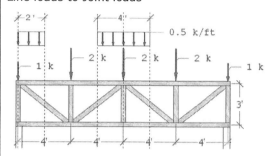   Edge joints : $\quad 0.5\ k/ft \times 2\ ft = 1\ kips$ Interior joints : $\quad 0.5\ k/ft \times 4\ ft = 2\ kips$

Truss Analysis (Method of Sections)
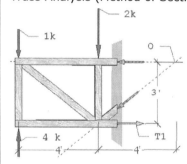 $\quad$ $\Sigma\ M_{(o)} = 0$ gives; $1\ k\ (8') + 2\ k\ (4') - 4\ k\ (8') + T1\ (4') = 0$ $T1\ (3') = -8\ k\text{-}ft - 8\ k\text{-}ft + 32\ k\text{-}ft$ $T1\ (3') = 16\ k\text{-}ft$ Therefore, $\quad T1 = 5.33\ kips = 5330\ lb$

Net Cross-sectional Area
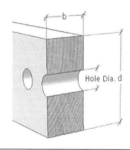 $\quad$ $A_n = A_g - A_h = b\,d - b\,(D_h) = b\,(d - D_h)$ $D_h = bolt\ diameter + \dfrac{1}{16} = \dfrac{3}{4} + \dfrac{1}{16} = 0.8125$ $A_n = 1.5" \times (5.5 - 0.8125) = 7.03\ in^2$

Actual Stress	$f_t = \dfrac{P}{A_n} = \dfrac{5330\ lb}{7.03\ in^2} = 758\ psi$
Allowable Stress	$F_t' = F_t\ (C_F)\ (C_D)\ (C_M)\ (C_t)\ (C_i)$ $= 400\ psi\ (1.3)\ (1.15)\ (1.0)\ (1.0)\ (1.0) = 598\ psi$
Decision	$f_t = 758\ psi > F_t' = 598\ psi \qquad$ Therefore, N. G. $\quad$ (Try 2x8)

Case Study 1b - Tension Member Design

Select the smallest dimension lumber size for the bottom chord of the truss shown.

1. Load combination, DL + SL, controls.

2. No ceiling attached to the bottom chord

3. Connections : Single row of 3/4-in.-diameter bolts

4. The uniform load transferred from the roof system to the top chords is 0.5 k/ft.

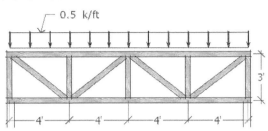

Line loads to Joint loads	
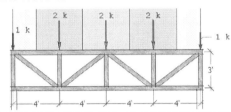	Edge joints : $\quad$ 0.5 k/ft x 2 ft = 1 kips Interior joints : $\quad$ 0.5 k/ft x 4 ft = 2 kips

Truss Analysis (Method of Sections)

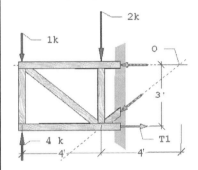

$\Sigma M_{(o)} = 0$ gives;

$$1 k\ (8') + 2 k\ (4') - 4 k\ (8') + T1\ (4') = 0$$

$$T1\ (3') = -8\ k\text{-}ft - 8\ k\text{-}ft + 32\ k\text{-}ft$$

$$T1\ (3') = 16\ k\text{-}ft$$

Therefore, $T1 = 5.33$ kips $= 5330$ lb

Allowable Stress

$$F_t' = F_t'\ (C_F)_{assumed}\ (C_D)\ (C_M)\ (C_t)\ (C_i)$$

$$= 400\ psi\ (1.3)(1.15)(1.0)(1.0)(1.0) = 598\ psi$$

Net Cross-sectional Area (required)

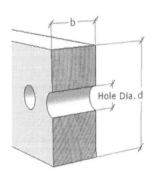

$$f_t = \frac{P}{A_n} = \frac{5330\ lb}{A_n} \leq F_t' = 598\ psi$$

$$A_n \geq \frac{P}{F_t'} = \frac{5330\ lb}{598\ \frac{lb}{in^2}} = 8.91\ in^2$$

Required $A_g = A_n + A_{hole} = 8.91\ in^2 + (\frac{3}{4} + \frac{1}{16})(1.5)\ in^2$

$$= 8.91\ in^2 + 1.22\ in^2 = 10.13\ in^2$$

From NDS section property table	Select $\quad$ 2 x 8 (11.25 in²) No. 1 S-P-F	$\dfrac{1.3}{1.2} < \dfrac{11.25}{10.13}$

C_F for 2 x 8 is 1.2 which is less than the assumed value, 1.3, **but** O.K. because.

Case study 2a - Tension Member Analysis

Determine the adequacy of the bottom (tension) chord of the truss.

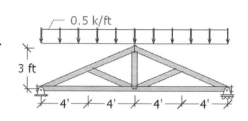

1. Load combination, DL + RLL, controls.

2. No ceiling attached to the bottom chord

3. Connections : Single row of 3/4-in.-diameter bolts

4. Bottom Chord : No. 1 Douglas Fir Larch, 2 x 8

5. The uniform load transferred from the roof system to the top chords is 0.5 k/ft.

Line loads to Joint loads

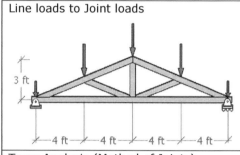

Edge joints :

Interior joints :

Truss Analysis (Method of Joints)

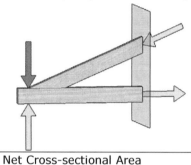

$\Sigma\, F_y = 0$ gives;

$\Sigma\, F_x = 0$ gives;

Net Cross-sectional Area

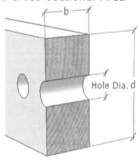

Hole Dia, d

$$A_n = A_g - A_h = b\, d - b\, (D_h) = b\, (d - D_h)$$

$$D_h = \text{bolt diameter} + \frac{1}{16} =$$

Actual Stress

$$f_t = \frac{P}{A_n} =$$

Allowable Stress

$$F_t' = F_t\, (C_F)\, (C_D)\, (C_M)\, (C_t)\, (C_i)$$

$$=$$

Decision

Case study 2b - Design

Determine the smallest dimension lumber size for the bottom chord of the truss shown.

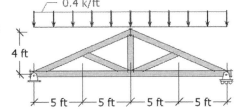

0.4 k/ft

4 ft

5 ft — 5 ft — 5 ft — 5 ft

1. Load combination, DL + SL, controls.
2. No ceiling attached to the bottom chord
3. Connections : Single row of 3/4-in.-diameter bolts
4. The uniform load transferred from the roof system to the top chords is 0.4 k/ft.

Line loads to Joint loads

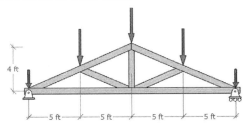

4 ft

5 ft — 5 ft — 5 ft — 5 ft

Edge joints :

Interior joints :

Truss Analysis (Method of Joints)

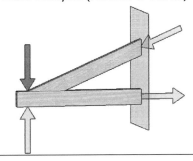

$\Sigma\, F_y = 0$ gives;

$\Sigma\, F_x = 0$ gives;

Allowable Stress

$F_t' = F_t'\,(C_F)_{assumed}\,(C_D)\,(C_M)\,(C_t)\,(C_i)$

$=$

Net Cross-sectional Area (required)

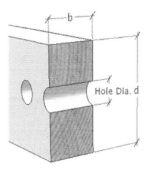

b

Hole Dia. d

$f_t = \dfrac{P}{A_n} = ($　　　　$) \le F_t' = ($　　　　$)$

$A_n \ge \dfrac{P}{F_t'} =$

Required $A_g = A_n + A_{hole} =$

Select the Lightest Section

Workshop 1a - Tension Member Analysis

Determine the adequacy of the bottom chord of the truss.

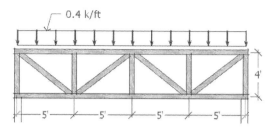

0.4 k/ft

1. Load combination, DL + RLL, controls.
2. No ceiling attached to the bottom chord
3. Connections : Single row of 3/4-in.-diameter bolts
4. Bottom Chord : No. 1 Douglas Fir Larch, 2 x 8
5. The uniform load transferred from the roof system to the top chords is 0.4 k/ft.

Line loads to Joint loads	
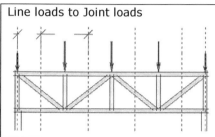	Edge joints : Interior joints :

Truss Analysis (Method of Sections)	
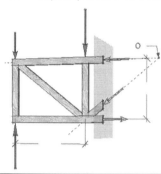	$\Sigma\, M_{(o)} = 0$ gives;

Net Cross-sectional Area	
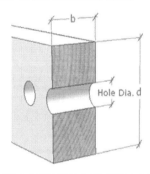	$A_n = A_g - A_h = b\,d - b\,(D_h) = b\,(d - D_h)$ $D_h = $ bolt diameter $+ \dfrac{1}{16} =$

Actual Stress	$f_t = \dfrac{P}{A_n} =$

Allowable Stress	$F_t' = F_t' \,(C_F)\,(C_D)\,(C_M)\,(C_t)\,(C_i)$ $=$

Decision	

Workshop 1b - Design

Determine the smallest dimension lumber size for the bottom chord of the truss shown.

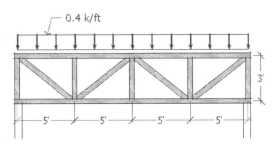
0.4 k/ft

3'

5' — 5' — 5' — 5'

1. Load combination, DL + SL, controls.

2. No ceiling attached to the bottom chord

3. Connections : Single row of 3/4-in.-diameter bolts

4. The uniform load transferred from the roof system to the top chords is 0.4 k/ft.

Line loads to Joint loads

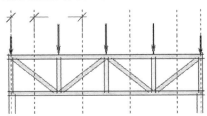

Edge joints :

Interior joints :

Truss Analysis (Method of Sections)

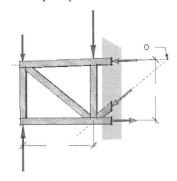

O

$$\Sigma M_{(o)} = 0 \text{ gives;}$$

Allowable Stress

$$F_t' = F_t' \ (C_F)_{assumed} \ (C_D) \ (C_M) \ (C_t) \ (C_i)$$

$$=$$

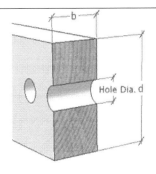

b

Hole Dia. d

Net Cross-sectional Area (required)

$$f_t = \frac{P}{A_n} = (\qquad) \le F_t' = (\qquad)$$

$$A_n \ge \frac{P}{F_t'} =$$

$$\text{Required } A_g = A_n + A_{hole} =$$

Select the Lightest Smallest Section

Workshop 2a - Tension Member Analysis

Determine the adequacy of the bottom (tension) chord in the truss shown below.

1. Load combination, DL + RLL, controls.
2. No ceiling attached to the bottom chord
3. Connections : Single row of 3/4-in.-diameter bolts
4. Bottom Chord : No. 1 Douglas Fir Larch, 2 x 8
5. The uniform load transferred from the roof system to the top chords is 0.5 k/ft.

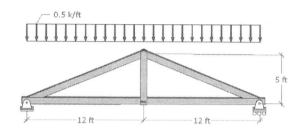

Workshop 2b - Design

Determine the smallest dimension lumber size for the bottom chord of the truss shown.

1. Load combination, DL + SL, controls.
2. No ceiling attached to the bottom chord
3. Connections : Single row of 3/4-in.-diameter bolts
4. The uniform load transferred from the roof system to the top chords is 0.4 k/ft.

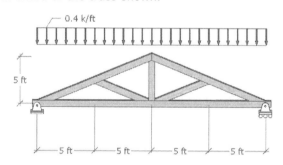

CHAPTER

4

BENDING STRESS

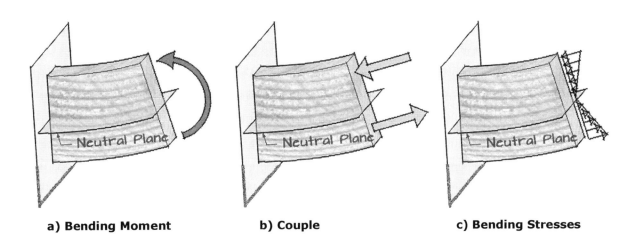

a) Bending Moment   b) Couple   c) Bending Stresses

4.1 Bending Moment and Couple

A bending moment can be resolved into two forces that are the same in magnitude and opposite in direction. The two forces are called a couple which is equivalent to the bending moment. The two forces in beam are one in compression and the other in tension. In reality, the two internal forces are not concentrated but distributed over the cross-section of the beam. These distributed internal forces can be viewed as stresses on the beam. The maximum compressive stress is found at the upper extreme fiber of the beam while the maximum tensile stress is located at the lower extreme fiber of the beam. Since the stresses between these two opposing maxima vary linearly, there exists a point on the linear path between them where there is no bending stress. The locus of these points is the neutral axis. Because of this area with no stress and the adjacent areas with low stress, using uniform cross section beams in bending is not a particularly efficient means of supporting a load as it does not use the full capacity of the beam until it is on the brink of collapse. Wide-flange beams (I-beams) and truss girders effectively address this inefficiency as they minimize the amount of material in this under-stressed region.

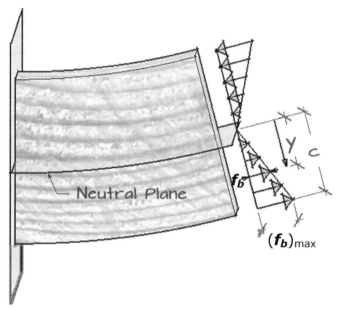

4.2 Elastic Bending Stress (fb)

The elastic bending stress varies linearly over the cross-section of the beam. This means that the stress distribution takes a triangular shape. The bending stress, f_b, at any point of the beam is obtained by:

$$f_b = \frac{M}{I} y$$

where, M = **Bending Moment** in the beam
I = **Moment of Inertia** of the cross-section
y = Distance from the Neural Plane

Note that the maximum tensile and compressive bending stresses acting at any given cross section occur at points farthest from the neutral axis. For a given cross section, M and I are constants and the only variable is y in the flexure formula. Thus, where the variable, y is the maximum, the bending stress, f_b, becomes the maximum.

4.3 Elastic Section Modulus (S)

To determine the maximum bending stress, put the maximum **y**-value, **c**, which is the distance from the neutral plane to the extreme fiber, in the bending stress formula. For a symmetrical section, the value of **c** is simply the half depth of the cross-section.

$$(f_b)_{max} = \frac{M}{I}c = \frac{M}{\left(\dfrac{I}{c}\right)} = \frac{M}{S}$$

where, **S = I/c = Elastic Section Modulus**

Therefore, if only the maximum bending stress is of interest (as is usual in structural design), the following simpler formula may be readily used.

$$\boxed{(f_b)_{max} = \frac{M}{S}}$$

Required Elastic Section Modulus

The bending design criteria for ASD can be expressed as:

$$\boxed{\textbf{Max. Actual Stress} \leq \textbf{Allowable Stress}}$$

If symbols are used, this inequality becomes:

$$(f_b)_{max} \leq F_b'$$

$$(f_b)_{max} = \frac{M}{S} \leq F_b'$$

$$\frac{M}{S} \leq F_b' \quad \text{or,} \quad S \geq \frac{M}{F_b'}$$

The ASD design criteria specify that the section modulus of the beam must be greater (or, at least equal) than $\dfrac{M}{F_b'}$ to avoid the bending failure. This is the minimum required value of section modulus and is used to select a trial beam size.

4.4 Graphical Study of Moment of Inertia

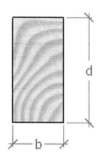

$$I = \frac{b\,d^3}{12}$$

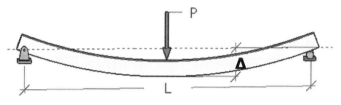

$$\Delta = PL^3 / 48\, EI$$

$b = 6$ in $\qquad d = 12$ in $I = bd^3/12 = 864$ in^4 $\Delta = PL^3/48\,EI$	
$b = 12$ in $\qquad d = 12$ in $I = bd^3/12 = 1728$ in^4 $\Delta = PL^3/48EI$	
$b = 6$ in $\qquad d = 24$ in $I = bd^3/12 = 6912$ in^4 $\Delta = PL^3/48E\,I$	

4.5 Graphical Study of Bending Stress

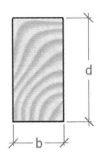

Max. Bending Stress
$f_b = \dfrac{M}{S}$

Section Modulus
$S = \dfrac{b\,d^2}{6}$

Thus, the **max. bending stress** becomes $\left(f_b\right)_{max} = \dfrac{6M}{bd^2}$

$M = 17280\ k\text{-}in, \qquad b = 6\ in, \quad d = 12\ in$ $f_b = 6M\,/\,bd^2$ $f_b = 6(17280)\,/(6)(12)^2$ $= 120\ ksi$	
$M = 17280\ k\text{-}in, \quad \boldsymbol{b = 12\ in}, \quad d = 12\ in$ $f_b = 6M\,/\,\mathbf{b}d^2$ $f_b = 6(17280)\,/\,\mathbf{12}(12)^2$ $= 60\ ksi$	
$M = 17280\ k\text{-}in, \quad b = 6\ in, \quad \boldsymbol{d = 24\ in}$ $f_b = 6M\,/\,b\mathbf{d^2}$ $f_b = 6(17280)\,/\,6(\mathbf{24})^2$ $= 30\ ksi$	

4.6 Workshop for Bending Stress

Case Study 4-1 Construct the load, shear and moment diagrams for the beam and calculate the maximum bending stresses (tensile and compressive bending stress).

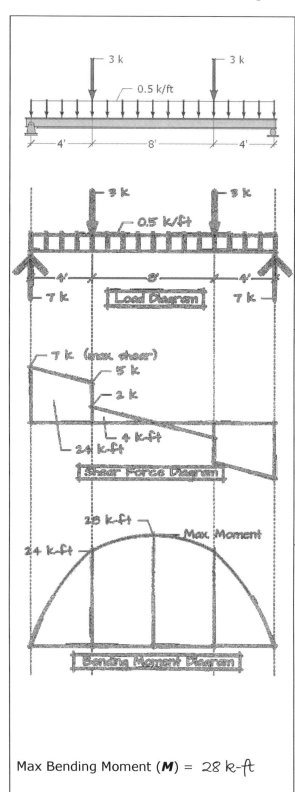

Max Bending Moment (**M**) = 28 k-ft

Cross-section

Nominal Size 6 x 20

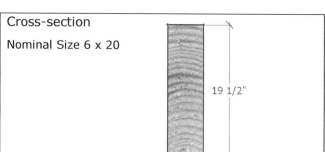

19 1/2"

5 1/2"

Moment of Inertia (I_x) =

$$I = \frac{b\,d^3}{12} = \frac{5.5\,(19.5)^3}{12} = 3399\ in^4$$

Section Modulus (S_x)

$$S_{tension} = \frac{I}{c_1} = \frac{3399\ in^4}{9.75\ in} = 349\ in^3$$

$$S_{comp} = \frac{I}{c_2} = \frac{3399\ in^4}{9.75\ in} = 349\ in^3$$

Tensile Bending Stress

$$(f_b)_{tension} = \frac{M_{max}}{S_{tension}} = \frac{28\,k-ft\,(12)}{349\ in^3}$$

$$= 0.962\ ksi = 962\ psi$$

Compressive Bending Stress

$$(f_b)_{comp} = \frac{M_{max}}{S_{comp}} = 962\ psi$$

Workshop 4-1a Construct the load, shear and moment diagrams for the beam and calculate the maximum bending stresses (tensile and compressive bending stress).

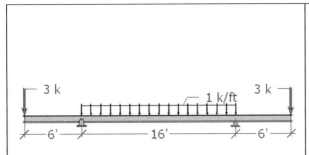

3 k   1 k/ft   3 k

6'   16'   6'

Max Bending Moment (**M**) =

Cross-section

Nominal Size 8 x 16

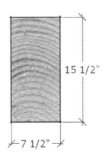

15 1/2"

7 1/2"

Moment of Inertia (I_x) =

$$I = \frac{b\,d^3}{12} =$$

Section Modulus (S_x)

$$S_{tension} = \frac{I_x}{c_1} =$$

$$S_{compression} = \frac{I_x}{c_2} =$$

Tensile Bending Stress

$$(f_b)_{tension} = \frac{M_{max}}{S_{tension}} =$$

Compressive Bending Stress

$$(f_b)_{comp} = \frac{M_{max}}{S_{comp}} =$$

Workshop 4-1b Construct the load, shear and moment diagrams for the beam and calculate the maximum bending stresses (tensile and compressive bending stress).

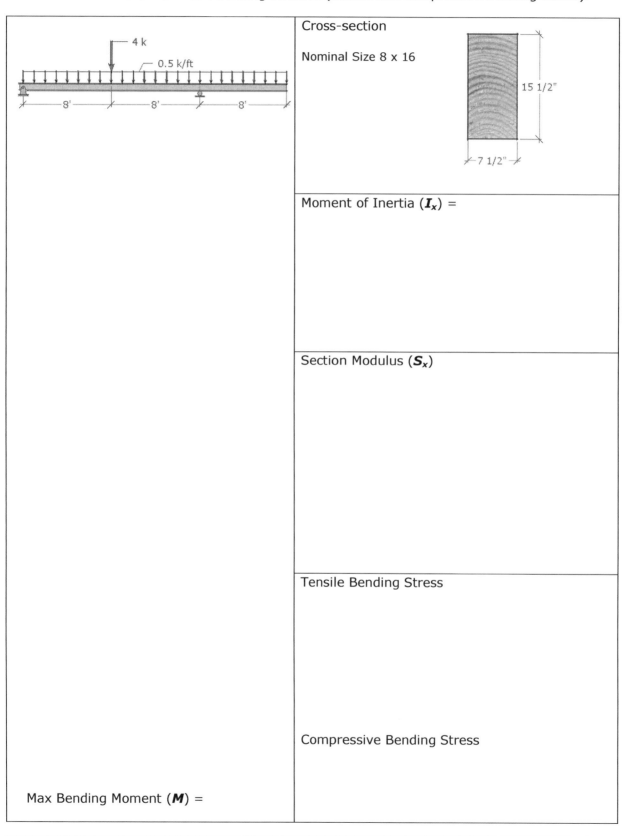

Cross-section

Nominal Size 8 x 16

15 1/2"

7 1/2"

Moment of Inertia (I_x) =

Section Modulus (S_x)

Tensile Bending Stress

Compressive Bending Stress

Max Bending Moment (**M**) =

Case Study 4-2 Determine the maximum bending tensile and compressive stresses for the beam given below.

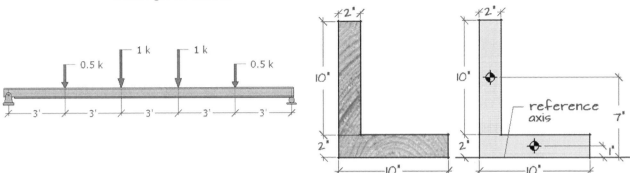

1. Locate *Neutral Axis*

Component	Area	y (from Centroid to Reference Axis)	A×y (Area Moment)
1	2 x10 = 20 in²	7 in	140 in³
2	10 x 2 = 20 in²	1 in	20 in³
ΣA =	40 in²	ΣA×y =	160 in³

$$\bar{y} = \frac{\sum(A \times y)}{\sum A} = \frac{160 \text{ in}^3}{40 \text{ in}^2} = 4 \text{ in}$$

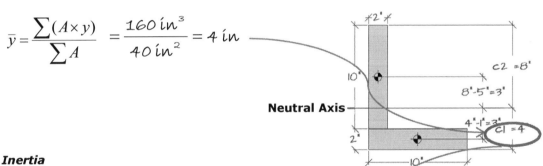

2. *Moment of Inertia*

Component	Original I_o	Bonus I		
	I_o	Area	$(d_y)^2$ (from centroid to Neutral Axis)	$A \cdot (d_y)^2$ (Moment of Inertia)
1	$\frac{2(10)^3}{12} = 166.7$	2x10 = 20 in²	(3 in)² = 9 in²	180 in⁴
2	$\frac{10(2)^3}{12} = 6.7$	2x10 = 20 in²	(3 in)² = 9 in²	180 in⁴
$\Sigma I_o =$	173.4 in⁴		$\Sigma A \cdot (d_y)^2 =$	360 in⁴

$$I_x = \Sigma I_o + \Sigma A \cdot (d_y)^2 = 173.4 \text{ in}^4 + 360 \text{ in}^4 = 533.4 \text{ in}^4$$

3. *Section Modulus*

$$S_{tension} = \frac{I_x}{c_1} = = \frac{533.4 \text{ in}^4}{4 \text{ in}} = 133.4 \text{ in}^3 \qquad S_{compression} = \frac{I_x}{c_2} = = \frac{533.4 \text{ in}^4}{8 \text{ in}} = 66.7 \text{ in}^3$$

4. Bending Moment Diagram

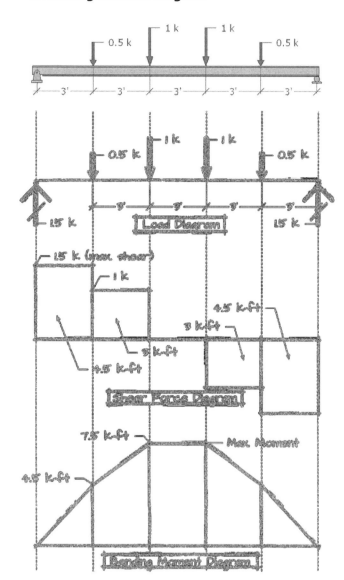

Max Bending Moment (*M*) = 7.5 k-ft

5. Bending Stresses

Tensile Bending Stress

$$(f_b)_{tension} = \frac{M_{max}}{S_{tension}} = \frac{7.5 \, k - ft (12)}{133.4 \, in^3} = 0.675 \ ksi = 675 \ psi$$

Compressive Bending Stress

$$(f_b)_{comp} = \frac{M_{max}}{S_{comp}} = \frac{7.5 \, k - ft (12)}{66.7 \, in^3} = 1.35 \ ksi = 1350 \ psi$$

Workshop 4-2a Determine the maximum bending tensile and compressive stresses for the beam given below.

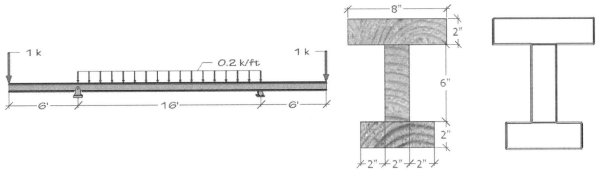

1. Locate *Neutral Axis*

Component	Area	y (*from Centroid to Reference Axis*)	A×y (*Area Moment*)

ΣA = [ ]   ΣA×y = [ ]

$$\bar{y} = \frac{\sum (A \times y)}{\sum A} =$$

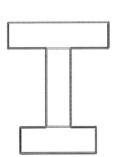

2. Moment of Inertia

Component	Original I_o I_o	Bonus I Area	$(d_y)^2$ (*from centroid to Neutral Axis*)	$A \cdot (d_y)^2$ (Moment of Inertia)

Σ I_o = [ ]   Σ $A \cdot (d_y)^2$ = [ ]

$$I_x = \Sigma I_o + \Sigma A \cdot (d_y)^2 =$$

3. Section Modulus

$$S_{tension} = \frac{I_x}{c_1} =$$

$$S_{compression} = \frac{I_x}{c_2} =$$

4. Bending Moment Diagram

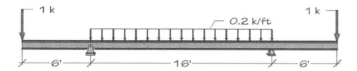

Max Bending Moment (*M*) =

5. Bending Stresses

Tensile Bending Stress

Compressive Bending Stress

1. Determine the maximum bending tensile and compressive stresses for the beam given below.

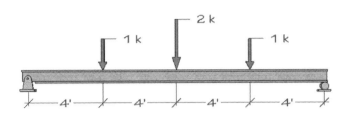

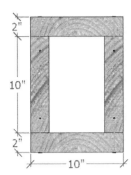

2. Determine the maximum bending tensile and compressive stresses for the beam given below.

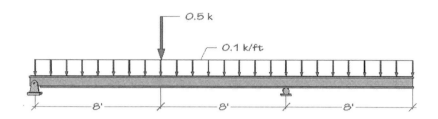

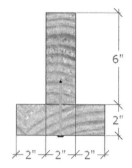

CHAPTER

5

SHEAR STRESS

5.1 Introduction

Shear stress arises from the force vector component parallel to the cross section. Normal stress, on the other hand, arises from the force vector component perpendicular to the material cross section on which it acts. Shear force (or stress) tends to slide the material on one side of a surface relative to the material on the other side of the surface in directions parallel to the surface without particular volume change. Opening a water bottle cap is a perfect example of shear force. Shear transferred directly from one member to another is known as direct shear and torsional shear is also developed in members due to torsion.

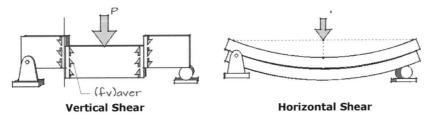

Vertical Shear   **Horizontal Shear**

Also, shear is generated in beams due to bending and is called bending shear. It is easy to imagine vertical (or, transverse) shear on a beam made of concrete blocks. Horizontal (or, longitudinal) shear can be demonstrated with a beam with multiple layers. Sliding occurs between layers if they are not fastened together. Let's further discuss these two bending shear stresses in the following sections.

5.2 Effects of Shear in Beams

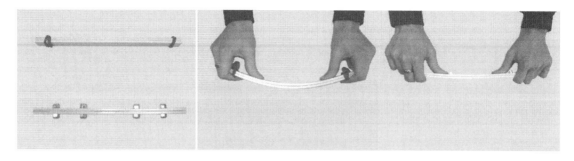

Two beam models   **Sliding between layers**   **No sliding between layers**
Less stiff beam   **Stiffer beam**

In the above figures, for one beam, the two plastic strips are loosely bound with elastic bands and for the other beam the two plastic strips are securely held together with four bolts that act as shear connectors and, because of the tension in the tightened bolts, provide compressive forces on the two strips. As shown, the bolts and the friction between the layers provide horizontal shear resistance and prevent the layers sliding between each other. This demonstrations show the existence of shear force in bending and how shear resistance can significantly increase the bending stiffness of a beam.

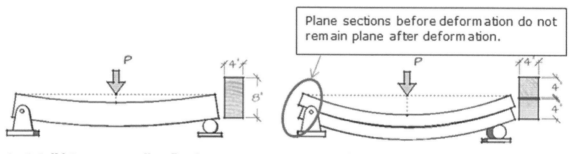

4 x8 Solid Beam - Small Deflection   **2-4 x4 Unfastened Beam - Large Deflection**

Now, consider a simple beam made up of two layers that are not bonded together and subjected to transverse loading. Each individual plank will be subjected to slip at the interfaces. Plane sections before deformation do not remain plane after deformation. If the planks are bonded together to form a single beam, longitudinal shear forces must develop to prevent the relative sliding between layers. This demonstraion indicates that there is a relation between shear force and bending moment.

5.3 Relationship between Bending and Shear

The shearing force at any section of a beam is the algebraic sum of the lateral components of the forces (including reactions) acting on either side of the section. Bending Moment at any action of a beam is the algebraic sum of the moments about the section of all forces acting on either side of the section. Thus both the shear force and bending moment of a specific section are functions of the loads on either side of the section. The following are some important relationships between shear force and bending moment diagrams:

1. The slope of the moment diagram at a given point is the shear at that point.
2. The area of the shear diagram to the section is equal to the moment at that section.
3. The maximum moment occurs at the point of zero shears. This is in reference to property number 2, that when the shear (also the slope of the moment diagram) is zero, the tangent drawn to the moment diagram is horizontal.
4. When the shear diagram is increasing, the moment diagram is concave upward.
5. When the shear diagram is decreasing, the moment diagram is concave downward.

5.4 Shear Force in Beams (Bending Shear)

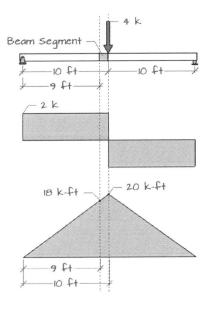

The first step in determining the bending shear force at any location in a beam is to examine a small beam segment. As shown in the bending moment diagram, the bending moments vary along this beam as usual. The bending moments at the left end and the right end of the beam segment are 18 k-ft and 20 k-ft respectively.

The bending moments at the left end of the beam segment is 18 k-ft, M_{b18} and 20 k-ft, M_{b20} at the right end. The associated bending stresses on the two cross-sections can be calculated using; $f_b = M/S$. Because this beam is prismatic (with a constant cross-section over the beam length), the section modulus, S, is constant. Thus the resulting bending stresses are directly proportional only to the bending moments, M. Therefore, f_{b18} ,the maximum tensile stress at the left end, is only 90% f_{b20}, the maximum tensile at the right end.

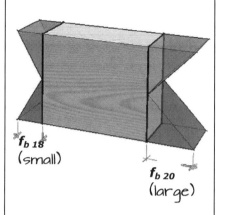

Let's take close look at a free-body taken from this beam segment. Either one will do but for simplicity, let's take the bottom piece.

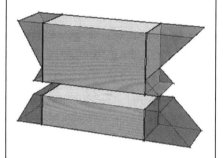

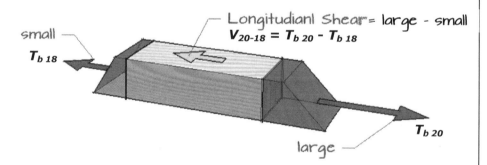

The resultants, T_{b18} and T_{b20} of the tensile bending stresses at the two cross-sections will be different and thus not in the horizontal equilibrium. To maintain the equilibrium of the beam piece, another horizontal force must exist. This is the longitudinal shear force, $V_{20-18} = T_{b20} - T_{b18}$, which is developed by the imbalance of the longitudinal bending stress on the horizontal section. Note that if the bending moments at both ends is the same (uniform bending moment), the longitudinal shear force in the segment is zero. This illustration proves the existence of the longitudinal shear stress.

5.5 Longitudinal and Transverse Shear

Considering equilibrium conditions of an infinitesimal particle at the cross-section a-a:

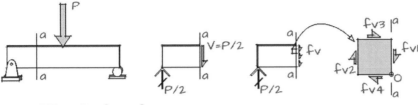

For $\Sigma F_x = 0$, $f_{v3} = f_{v4}$

For $\Sigma F_y = 0$, $f_{v1} = f_{v2}$

For $\Sigma M_o = 0$, $f_{v1} = f_{v4}$ and $f_{v2} = f_{v3}$

When a shear stress is exerted on a vertical face of a particle, another shear stress of the same magnitude must be exerted in the opposite direction on the other vertical face of the particle for the vertical equilibrium. Also, another pair of shear stresses must exist on the horizontal faces for the rotational equilibrium.

Thus, it is concluded that *the magnitudes of the longitudinal shear and transverse shear are the same*.

Horizontal Shear Failure of Lumber

Even though shear stress is usually not the primary beam design concern, because the longitudinal (or, horizontal) shear is the most critical in wood beam design, actual horizontal shear, f_v, must always be checked against the allowable shear stress, F_v'. Short span beams subjected to large concentrated loads are expected to fail due to shear, not bending. In wood beams, horizontal shear failure will always occur before vertical shear failure. Thus, it is not necessary to check for vertical shear except for beams notched at their supports.

5.6 Shear Stress Distribution

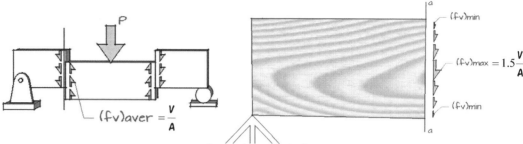

The real Vertical Shear Stress varies over cross-section

The real shear stress on a cross-section of a beam is not uniformly distributed but non-linearly distributed as calculated by:

$$f_v = \frac{V Q}{I b}$$

where, f_v = shear stress
V = shear force
Q = first moment of area above plain of interest
I = moment of inertia about neutral axis
b = width of cross-section

Since the shear stress, f_v is a function of Q, the maximum shear stress takes place at the neutral axis location where Q-value is the maximum because the largest area will be obtained above the neutral axis. f_v will be zero at the top and bottom (because $Q = 0$ at top/bottom) of the cross section.

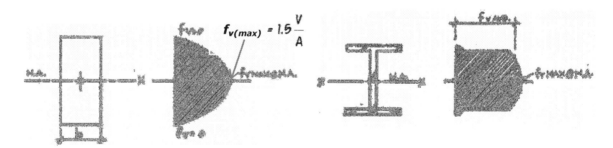

For a **rectangular cross-section**, the maximum shear stress can be calculated by using a simplified formula:

$$f_{v(max)} = 1.5\frac{V}{A} = 1.5 \times f_{v(average)}$$

5.7 Workshop for **Shear Stress**

Case Study 5-1

Determine the shear stresses at the points **b, c** and **d** and the maximum shear stress of the cross section a-a of a glued laminated beam as shown below.

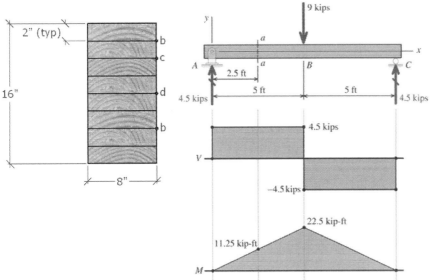

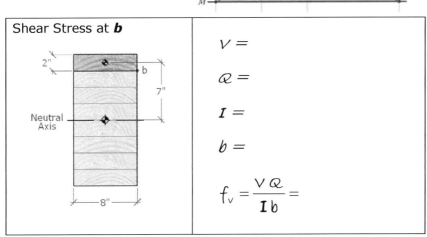

Shear Stress at **b**	
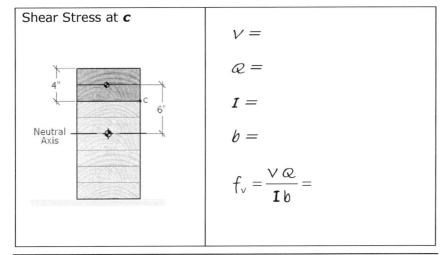	$V =$ $Q =$ $I =$ $b =$ $f_v = \dfrac{V\,Q}{I\,b} =$

Shear Stress at **c**	
	$V =$ $Q =$ $I =$ $b =$ $f_v = \dfrac{V\,Q}{I\,b} =$

Shear Stress at **d**	
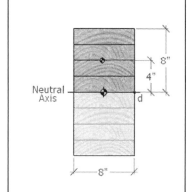	$V =$ $Q =$ $I =$ $b =$ $f_v = \dfrac{V Q}{I b} =$

Shear Stress at **e**	
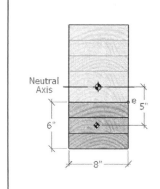	$V =$ $Q =$ $I =$ $b =$ $f_v = \dfrac{V Q}{I b} =$

Shear Stress Distribution along Depth of Rectangular Beam

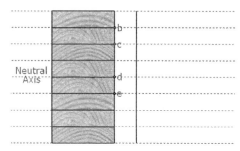

Discussion

1. Where does the maximum shear stress take place?

2. What is the maximum shear stress?

3. What is the average shear stress, $(f_v)_{average}$ = **V/A**?

4. **What is the ratio of the maximum to the average shear stresses?**

Workshop 5-1a
The following cross section is subjected to a shear force of 1.3 kips.
1. Determine the shear stress at **A**.
2. Determine the maximum shear stress of the cross section.

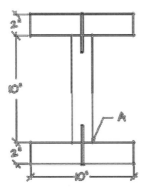

Workshop 5-1b
The following cross section is subjected to a shear force of 1.5 kips.
1. Determine the shear stress at **A**.
2. Determine the maximum shear stress of the cross section.

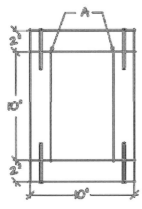

Workshop 5-1c
The following cross section is subjected to a shear force of 1.0 kips.
1. Determine the shear stresses at **A**.
2. Determine the maximum shear stress of the cross section.

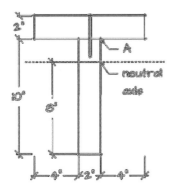

Case Study 5-2

The shear capacity of the nail used for the built-up beam is 350 lb. The maximum shear force in the beam is 300 lb. Check the adequacy of the nailed connection.

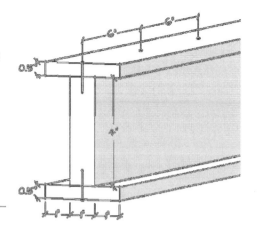

Solution)

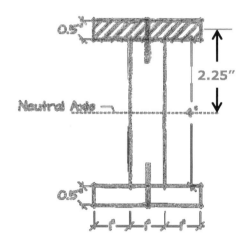

$$f_v = \frac{V Q}{I b}$$

$V = 300 \ lb \qquad b = 1 \ in$

Moment of Inertia

$$I = \frac{3(5)^3}{12} - 2 \times \frac{1(4)^3}{12} = 20.58 \ in^4$$

$$Q = (3 \times 0.5) \times (2.25) = 3.375 \ in^3$$

$$f_v = \frac{(300 \ lb)(3.375 \ in^3)}{(20.58 \ in^4)(1 \ in)} = 49.2 \frac{lb}{in^2}$$

Nail Force

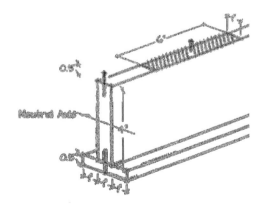

Stressed Area

$A = (6 \ in) \times (1 \ in) = 6 \ in^2$

Nail Force (Shear force in Nail)

$$V = f_v \times A = 49.2 \frac{lb}{in^2} \times (6 \ in) = 295 \ lb$$

Decision

Actual Nail Force = 295 lb < Allowable Nail Strength = 350 lb (O.K. or N.G.)

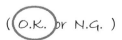

Cast Study 5-3a

The shear capacity of the nail used for the built-up beam is 350 lb. The maximum shear force in the beam is 300 lb. Check the adequacy of the nail connection.

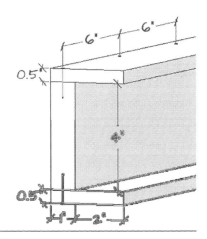

Solution)

$$f_v = \frac{V Q}{I b}$$

$$V = (\qquad) \; lb \qquad\qquad b = (\qquad) \; in$$

Moment of Inertia

$$I =$$

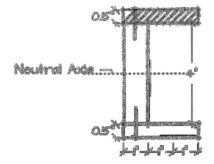

$$Q =$$

$$f_v = \frac{V Q}{I b} =$$

Nail Force

Stressed Area

$$A =$$

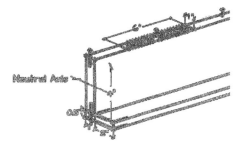

Nail Force (Shear force in Nail)

$$V = f_v \times A =$$

Decision

Cast Study 5-3b

The shear capacity of the nail used for the built-up beam is 250 lb. The maximum shear force in the beam is 400 lb. Check the adequacy of the nail connection.

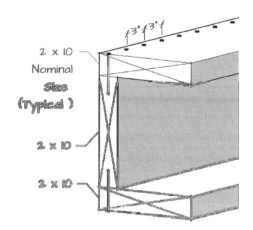

2 x 10
Nominal
Size
(Typical)

2 x 10

2 x 10

Solution)

$$f_v = \frac{V Q}{I b}$$

$$V = (\qquad) \, lb \qquad\qquad b = (\qquad) \, in$$

Moment of Inertia

$$I =$$

$$Q =$$

$$f_v = \frac{V Q}{I b} =$$

Nail Force

Stressed Area

$$A =$$

Nail Force (Shear force in Nail)

$$V = f_v \times A =$$

Decision

Case Study 5-4

The shear capacity of the nail used for the built-up beam is 250 lb. The maximum shear force in the beam is 800 lb. Check the adequacy of the nail connection. The nail spacing is 3 in. o.c.

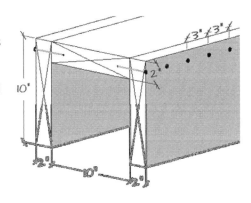

Locate Neutral Axis

Component	Area	y (from Centroid to Reference Axis)	$A \times y$ (Area Moment)
$\Sigma A =$ [ ]		$\Sigma A \times y =$ [ ]	

$$\overline{y} = \frac{\sum (A \times y)}{\sum A} =$$

Moment of Inertia

Component	Original I_o		Bonus I	
	I_o	Area	$(d_y)^2$ (from centroid to Neutral Axis)	$A \cdot (d_y)^2$ (Moment of Inertia)
$\Sigma I_o =$ [ ]			$\Sigma A \cdot (d_y)^2 =$ [ ]	

$$I_x = \Sigma I_o + \Sigma A \cdot (d_y)^2 =$$

$$f_v = \frac{V \, Q}{I \, b}$$

$V = ($ $)$ lb $b = ($ $)$ in

Q-value

$$Q =$$

$$f_v = \frac{V \, Q}{I \, b} =$$

Nail Force

Stressed Area

$$A =$$

Nail Force (Shear force in Nail)

$$V = f_v \times A =$$

Workshop 2

3-2 in by 10 in boards are nailed together to form a channel shape beam. The shear capacity of the nail used for the built-up beam is 250 lb. The maximum shear force in the beam is 800 lb. Check the adequacy of the nail connection. The nail spacing is 3 in. o.c.

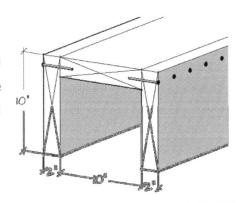

Locate Neutral Axis

Component	Area	y (from Centroid to Reference Axis)	$A \times y$ (Area Moment)
$\Sigma A =$ [ ]		$\Sigma A \times y =$ [ ]	

$$\bar{y} = \frac{\sum(A \times y)}{\sum A} =$$

Moment of Inertia

Component	Original I_o		Bonus I	
	I_o	Area	$(d_y)^2$ (from centroid to Neutral Axis)	$A \cdot (d_y)^2$ (Moment of Inertia)
$\Sigma I_o =$ [ ]			$\Sigma A \cdot (d_y)^2 =$ [ ]	

$$I_x = \Sigma I_o + \Sigma A \cdot (d_y)^2 =$$

$$f_v = \frac{V Q}{I b}$$

$$V = (\qquad) \; lb \qquad b = (\qquad) \; in$$

Q-value

$$Q =$$

$$f_v = \frac{V Q}{I b} =$$

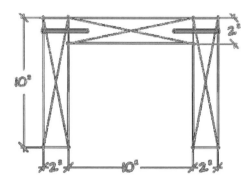

Nail Force

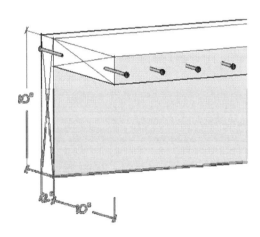

Stressed Area

$$A =$$

Nail Force (Shear force in Nail)

$$V = f_v \times A =$$

Workshop 3

3-2x10 nominal size members are nailed together to form a channel shape beam. The shear capacity of the nail used for the built-up beam is 250 lb. The maximum shear force in the beam is 900 lb. Check the adequacy of the nail connection.

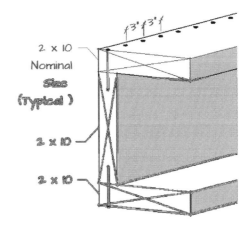

2 x 10
Nominal
Size
(Typical)

2 x 10

2 x 10

Workshop 4

2-2x10 actual size members are nailed together to form a T-shape beam. The shear capacity of the nail used for the built-up beam is 250 lb. The maximum shear force in the beam is 900 lb.

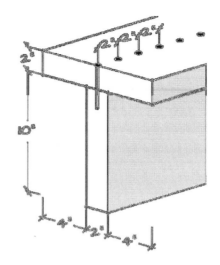

CHAPTER

6

BEAM DESIGN

6.1 Beam Analysis and Design Processes

Structural analysis is a process to determine if an existing member (or structure) is safe or unsafe with the given member size, loading, material and geometry. The beam analysis process includes the following:

1. Find the maximum shear force and bending moment.
2. Determine the resulting actual stresses.
3. Determine the allowable stresses.
4. Check that the actual stresses are less or equal to the allowable stresses.
5. Check that the deflections are less than the code specified values.
6. Check that the bearing stresses at support locations are less than the code specified values.

Structural design involves the arrangement and proportioning of new structures and their components in such a way that the assembled structure is capable of supporting the designed loads within the allowable limit states. Similarly, wooden beam design is basically an iterative trial-and-error process because the beam size is not known thus the beam weight and size factor must be assumed at the beginning. With the guessed values, the required section modulus for the given loads is determined. And then, a trial size for the beam is selected based on the required bending capacity because bending is the primary controlling factor for beams. The selected beam is analyzed for the bending stress, shear stress and deflection limitations. The beam design process includes the following:

1. Find the maximum shear force and bending moment.

2. Determine the allowable stresses.

3. Determine the required section modulus.

4. Select the lightest trial section with the required section modulus.

5. Check the shear stress and deflections.

6. Check that the bearing stresses at support locations are less than the code specified values.

6.2 Bearing Stress (compression perpendicular to grain)

The final step of beam analysis/design process is to make sure the selected lumber for the applied loads has the required design strength for the compression perpendicular to the grain. The loads carried by a joist, beam or similar wood member are transferred through their ends to supporting walls, beams or columns. The reaction

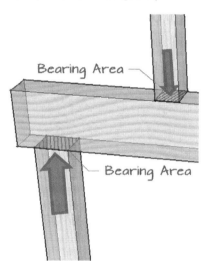

from the supporting member tends to compress the fibers of the beam in perpendicular to the grain. It has been emphasized in Chapter 1 that wood has extremely low compressive strength in the direction perpendicular grain compared to compression parallel to grain. The bearing area must be sufficient in size to prevent crushing (e.g., a sill plate with studs bearing down on it).

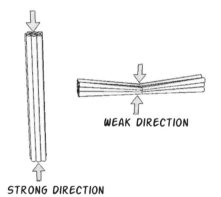

Bearing stress, along with bending and shear, must be investigated for beams – particularly short beams with very heavy loads. This type of stress makes a beam also very susceptible to crushing if the MC is greater than 19% and also the bearing areas of supports are relatively small. The allowable compressive stress perpendicular to grain, $F'_{c\perp}$ shall be greater than the actual compressive stress perpendicular to grain, $f_{c\perp}$. For sawn lumber, the Reference Design Values for compressive perpendicular to grain, $F_{c\perp}$, are based on a deformation limit that has been found by experience to provide adequate service in typical wood frame

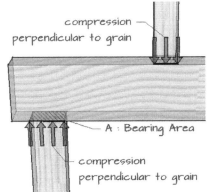

construction. Therefore, load duration factor is not applicable to compressive allowable stress perpendicular to grain.

$$f_{c\perp} = \frac{V}{A} \leq F'_{c\perp}$$

where:

$f_{c\perp}$ = actual compressive stress perpendicular to grain

V = maximum vertical shear a at point of interest

A = bearing area or, stressed area in^2

$F'_{c\perp} = F_{c\perp} \times (C_M) \times (C_t) \times (C_i) \times (C_b)$

$F_{c\perp}$ = tabulated allowable stress values from table above

C_M = wet-service factor

C_t = load duration factor

C_b = bearing area factor used if bearing length, $L_b \leq 6"$ and not nearer than 3" from the end of the member

It must be also noted that a force generating tension perpendicular to grain tends to separate the wood fibers along the grain. This is the direction in which wood has the least strength. Thus, it is not good practice to apply loading to induce tension across grain. For this reason, design values are not provided for this strength property.

6.3 Sloped Beam Analysis

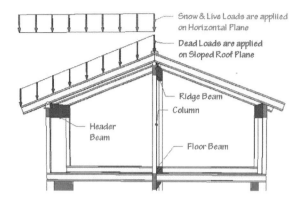

For sloped beams such as rafters in a sloped roof or stair stringer, the effects of the slope on the behavior of the beam have to be correctly addressed.

Per IBC 1607.11, both the snow and live loads act vertically on the horizontal projection while the dead load the acts on the sloped plane. To add these loads together, they must be consistent - either along the horizontal plane or along the slope. Because sloped beams are found in many architectural applications, it is prudent to review the comparison of the two different analysis methods, the *horizontal plane method* and the *sloped roof plane method*. The resulting addition of loads with respect to the two different planes is given in the two analysis method.

With the *horizontal plane method*, an increased uniform load is applied directly to a hypothetical beam with the horizontally projected span of the inclined actual beam. The horizontal plane method is used widely due to its simplicity and relative accuracy. However, the horizontal thrust at the left pin support must be considered separately as to be addressed in the sloped roof plane method.

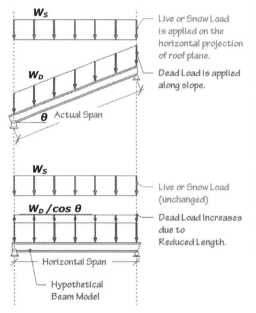

With the *sloped roof plane method*, the uniform load is resolved into components of load perpendicular to the longitudinal axis of the beam (bending) and parallel to the longitudinal axis of the beam (compression), and the span length is the same as the inclined actual span. The horizontal tractions do not cause any

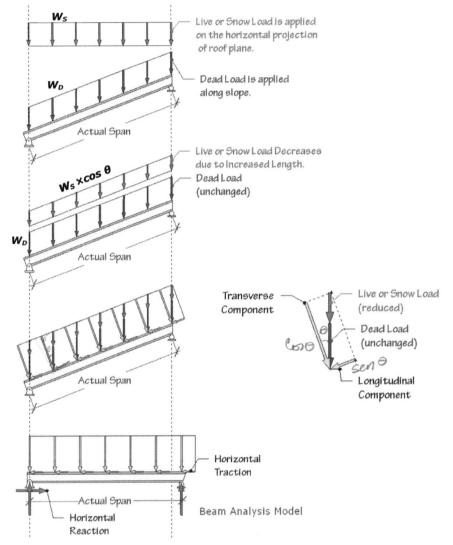

Beam Analysis Model

bending moment of the beam but will increase the axial stresses on the compression zone of the beam cross-sections. Note that this horizontal reaction of the beam in turn pushes the supporting wall outward. It has been shown that the two methods result in very similar bending moment and shear values, but also that the axial compression portion of the sloped roof plane method is insignificant when considering the interaction of compression and bending.

6.4 Workshop for **Bearing Area Design**
Case Study 6-1a

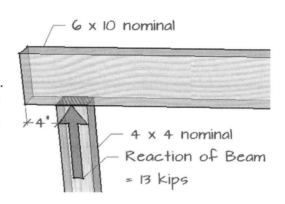

6 x 10 nominal

4"

4 x 4 nominal

Reaction of Beam = 13 kips

The nominal 6 x 10 Douglas Fir Larch No. 1 wood beam is sitting on a nominal 4 x 4 column as shown. The reaction of the beam is 13 kips. Determine if the beam is acceptable for compression perpendicular to grain.

Solution)

$$f_{c\perp} = \frac{V}{A} = \frac{10000\ lb}{(5.5\ in)(3.5\ in)} = 675\ \frac{lb}{in^2}$$

Since the distance from the end of the beam is 5" which is greater than 3" AND the bearing length, $L_b = 3.5"$ (actual dimension of 4 x 4), C_b may be greater than 1.0.

$$C_b = \frac{L_b + 0.375}{L_b} = \frac{3.5 + 0.375}{3.5} = 1.11$$

$$F'_{c\perp} = F_{c\perp} \times (C_M) \times (C_t) \times (C_i) \times (C_b) = (625\ psi) \times (1.0) \times (1.0) \times (1.0) \times (1.11) = 692\ psi$$

$$F_{c\perp} = 675\ psi \leq F'_{c\perp} = 692\ psi\ \ O.K.$$

Case Study 6-1b

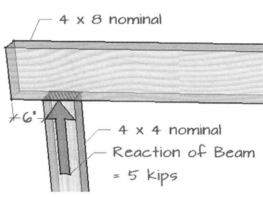

4 x 8 nominal

6"

4 x 4 nominal

Reaction of Beam = 5 kips

The nominal 4 x 8 Hem Fir No. 1 wood beam is sitting on a nominal 4 x 4 column as shown. The reaction of the beam is 5 kips. Determine if the beam is acceptable for compression perpendicular to grain.

Solution)

$$f_{c\perp} = \frac{V}{A} =$$

$$C_b = \frac{L_b + 0.375}{L_b} = \frac{(\qquad) + 0.375}{(\qquad)} = (\qquad)$$

$$F'_{c\perp} = \quad F_{c\perp} \quad \times \quad (C_M) \quad \times \quad (C_t) \quad \times \quad (C_i) \quad \times \quad (C_b)$$

$$= (\qquad) \times (\qquad) \times (\qquad) \times (\qquad) \times (\qquad) =$$

$$F_{c\perp} = (\qquad) \leq F'_{c\perp} = (\qquad). \quad (\ O.K. \quad or \quad N.G.\ )$$

Workshop 6-1a

The nominal 4 x 10 Douglas Fir Larch Select Structural wood beam is sitting on a nominal 6 x 4 column as shown. The reaction of the beam is 12 kips. Determine if the bearing area of the beam is safe for the compressive stress perpendicular to grain.

4x4 nominal — Column Load = 8 kips
4x10 nominal
Reaction of Beam = 12 kips
6x4 nominal
4"
5.5"

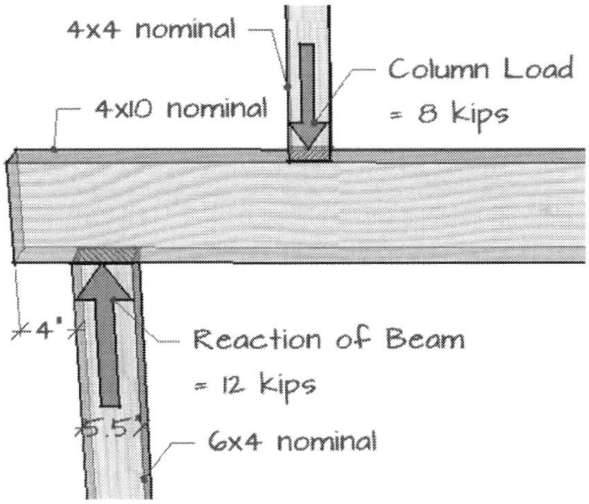

Solution)

Workshop 6-1b

The above beam in Workshop 1 supports a nominal 4 x 4 column as shown. The column load is 8 kips. Determine if the bearing area on the beam is safe for the compressive stress perpendicular to grain.

Solution)

Workshop 6-1c

Determine the adequacy of the truss support detail based on the compressive stress perpendicular to grain of the Hem-Fir No. 2 nominal 2x4 bottom chord of the truss. The reaction of the truss is 3000 lb.

Solution)

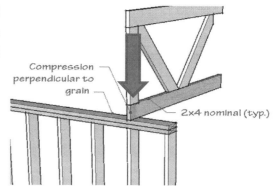

Workshop 6-1b

Determine the adequacy of the truss support detail based on the compressive stress parallel to grain of the Hem-Fir No. 2 nominal 2x4 stud. The reaction of the truss is 3000 lb.

Solution)

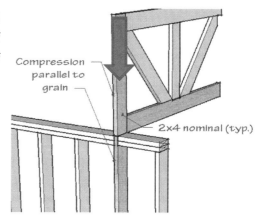

6.5 Workshop for **Selection of Trial Beam Size**

Case Study 6-2a Select the lightest section based on the section modulus required.

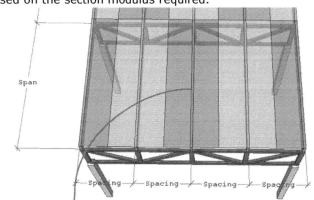

Design Data :

    Joist Span = 18 ft.

    Joist Spacing = 16 in.

    D.L. = 10 lb/ft^2

    L.L. = 50 lb/ft^2

    Douglas-Fir Larch, No 1 Grade

    $F_b{'}$ = 1035 psi

From Area Load to Line Load

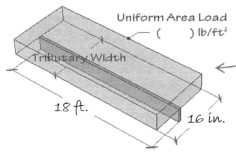

$$D.L. = 10\,\frac{lb}{ft^2} \times \left(\frac{16}{12}\right)ft = 13.33\,\frac{lb}{ft}$$

$$L.L. = 50\,\frac{lb}{ft^2} \times \left(\frac{16}{12}\right)ft = 66.7\,\frac{lb}{ft}$$

T.L. =   80 plf

Bending Moment

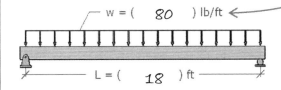

w = (80) lb/ft

L = (18) ft

$$M_{max} = \frac{w\,L^2}{8} = \frac{\left(80\,\frac{lb}{ft}\right)(18\,ft)^2}{8} = 3240\,lb-ft$$

Required Section Modulus

$$\left(f_b\right)_{max} = \frac{M_{max}}{S_x} \le F_t{'} \qquad or, \qquad S_x \ge \frac{M_{max}}{F_t{'}}$$

$$S_x \ge \frac{3240\,lb-ft\,(12\,\frac{in}{ft})}{1035\,\frac{lb}{in^2}} = 37.57\,in^3$$

Select Lightest TRIAL Size

	S_x	A	
2 x 14	43.89 in³	19.88 in²	<-- Try the smallest cross-section
3 x 12	52.73 in³	28.12 in²	
4 x 10	49.91 in³	32.38 in²	

Case Study 6-2b Select the lightest section for the joist based on the section modulus required.

Design Data :

Joist Span = 19 ft.

Joist Spacing = 18 in.

D.L. = 10 lb/ft^2

L.L. = 20 lb/ft^2

Douglas-Fir Larch, No 1 Grade

F_b' = 1005 psi

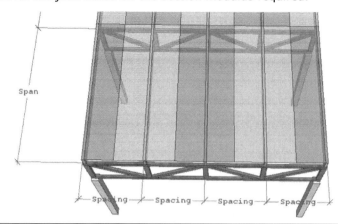

From Area Load to Line Load

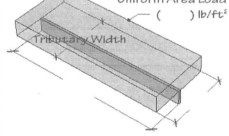

Bending Moment

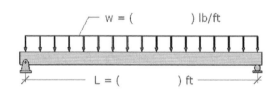

Required Section Modulus

Select Lightest TRIAL Size

Workshop 6-2a Select the lightest section for the joist based on the section modulus required.

Design Data :

Joist Span = 14 ft.

Joist Spacing = 24 in.

D.L. = 15 lb/ft^2

L.L. = 25 lb/ft^2

F_b' = 1100 psi

Hem-Fir, No 2 Grade

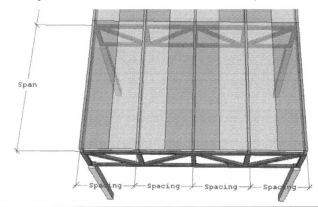

From Area Load to Line Load
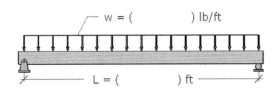
Bending Moment
w = ( ) lb/ft L = ( ) ft
Required Section Modulus
Select Lightest _TRIAL_ Size

Workshop 6-2b Select the lightest section for the joist based on the section modulus required.

Design Data :

    Joist Span = 16 ft.

    Joist Spacing = 18 in.

    D.L. = 14 lb/ft^2

    L.L. = 25 lb/ft^2

    F_b' = 1100 psi

    Douglas-Fir South, No 2 Grade

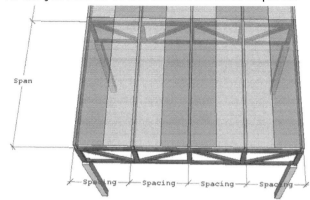

From Area Load to Line Load
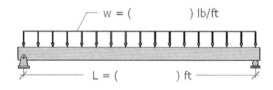
Bending Moment
w = (   ) lb/ft L = (   ) ft
Required Section Modulus
Select Lightest TRIAL Size

Extra Exercise Calculate the *total load deflection* and the *live load deflection* of each member below.

1. Design Data :   a. Douglas Fir-Larch No. 1 Grade
                    b. 2×10 roof joists span 20 feet.
                    c. and support a *DL* of 20 lb/ft and a *RLL* of 40 lb/ft

$$\Delta_{total} = \frac{5\, W_{total}\, L^4\, (12^3)}{384\, EI} = \frac{5(20+40)\ lb/ft\ (20\ ft)^4 (12)^3}{384\,(1700000)\, lb/in^2\, (98.93\ in^4)} = 1.28\ in.$$

$$\Delta_{live} = \Delta_{total} \times \frac{W_{live}}{W_{total}} = 1.28\ in \times \frac{40}{60} = 0.86\ in.$$

2. Design Data :   a. Douglas Fir-Larch No. 2 Grade
                    b. 2×10 roof joists span 17 feet
                    c. and support a *DL* of 30 lb/ft and a *SL* of 60 lb/ft.

3. Design Data :   a. Douglas Fir-Larch No. 1 Grade
                    b. 2×12 floor joists span 20 feet
                    c. and support a *DL* of 28 lb/ft and a *LL* of 80 lb/ft .

4. Design Data :   a. Hem-Fir No. 1 Grade
                    b. 3×16 floor beams span 22 feet
                    c. and support a *DL* of 70 lb/ft and a *LL* of 200 lb/ft.

5. Design Data :   a. Douglas Fir-Larch No. 1 Grade
                    b. 6×16 roof beams span 16 feet
                    c. and support a *DL* of 225 lb/ft and a *RLL* of 525 lb/ft.

6.6   Workshop for **Beam Design**
Case Study 6-3
    Design Data
        Dead Load   : 15 psf
        Snow Load   : 35 psf

        Roof deflection limit   : $\Delta_T \leq$ L/180 and $\Delta_L \leq$ L/240.

a. Select the most economical size for the joist, J1, using Douglas Fir South, No 2.

    J1 : _____

b. Select the most economical for the beam, B2, using Douglas Fir Larch, No1.

    B2 : _____

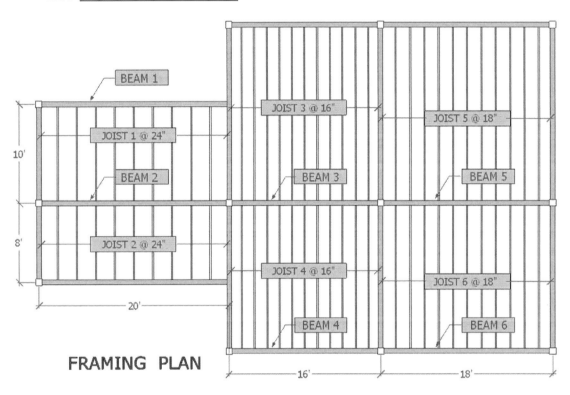

FRAMING PLAN

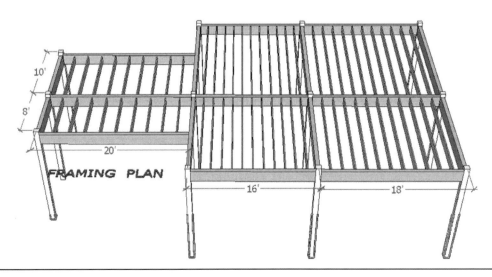

Case Study 6-3a Select the lightest section based on the section modulus required.

Design Data for (_Joist 1_) :

Joist or Beam Span = (10 ft)				Tributary Width = (24 in)	

Loads : D.L. = 15 psf	S.L. = 35 psf	

Species : Douglas Fir South	Grade : No 2

Allowable Stresses :

	Reference Design Values	(C_F) assumed	C_D	C_r	Adjusted Value
$F_b' =$	850 psi	1.1	1.15	1.15	1237 psi
$F_v' =$	180 psi	N/A	1.15	N/A	207 psi
$E' =$	1200 ksi	N/A	N/A	N/A	1200 ksi

From Area Load to Line Load

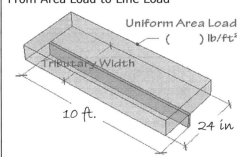

Uniform Area Load
( ) lb/ft²

Tributary Width

10 ft.

24 in

$$D.L. = 15\frac{lb}{ft^2} \times \left(\frac{24}{12}\right)ft = 30\frac{lb}{ft}$$

$$L.L. = 35\frac{lb}{ft^2} \times \left(\frac{24}{12}\right)ft = 70\frac{lb}{ft}$$

$$T.L. = \qquad 100 \text{ plf}$$

Bending Moment

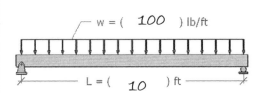

W = (100) lb/ft

L = (10) ft

$$M_{max} = \frac{wL^2}{8} = \frac{\left(100\frac{lb}{ft}\right)(10\,ft)^2}{8} = 1250 \text{ lb-ft}$$

Required Section Modulus

$$\left(f_b\right)_{max} = \frac{M_{max}}{S_x} \le F_t' \qquad \text{or,} \quad S_x \ge \frac{M_{max}}{F_t'}$$

$$S_x \ge \frac{1250\,lb\text{-}ft\,(12\frac{in}{ft})}{1237\frac{lb}{in^2}} = 12.13 \text{ in}^3$$

Select Lightest _TRIAL_ Size

	S_x	A	
2 x 8	13.14 in³	10.88 in²	<-- Try the smallest cross-section
3 x 6	12.60 in³	13.75 in²	
4 x 6	17.65 in³	19.25 in²	

Trial Size (2 X 8)

$S_x = 13.14 \text{ in}^3$	$I_x = 47.63 \text{ in}^4$	$A = 10.88 \text{ in}^2$

Actual	Allowable
Bending Stress $(f_b)_{max} = \dfrac{M_{max}}{S}$ $= \dfrac{1250 \text{ lb-ft} \left(12 \dfrac{in}{ft}\right)}{13.14 \text{ in}^3}$ $= 1142 \text{ psi}$	(note: C_F for 2 x 8 must be used.) |	C_F	C_D	C_r	| |---|---|---|---|---| | $F_b' =$ | 1350 | 1.2 | 1.15 | 1.15 | 1349 psi | **Design Criteria** $(f_b)_{max} \leq F_t'$ 1142 psi < 1349 psi O.K.
Shear Stress $V_{max} = \dfrac{WL}{2} = \dfrac{\left(100 \dfrac{lb}{ft}\right)(10 \text{ ft})}{2} = 500 \text{ lb}$ $(f_v)_{max} = 1.5 \dfrac{V_{max}}{A}$ $= 1.5 \dfrac{500 \text{ lb}}{10.88 \text{ in}^2} = 69 \text{ psi}$			C_F	C_D	C_r	| |---|---|---|---|---| | $F_v' =$ | 180 | N/A | 1.15 | N/A | 207 psi | **Design Criteria** $(f_v)_{max} \leq F_v'$ 69 psi < 207 psi O.K.
Deflection $\Delta_{total} = \dfrac{5 \, W_{total} \, L^4 (12^3)}{384 \, E \, I}$ $= \dfrac{5(100)(10)^4(12^3)}{384(1200000)(47.63)} = 0.39$ in. $\Delta_{live} = \Delta_{total} \times \dfrac{W_{live}}{W_{total}} = 0.39 \text{ in.} \times \dfrac{70}{100}$ $= 0.27 \text{ in.}$			C_F	C_D	C_r	| |---|---|---|---|---| | $E' =$ | 1400 | N/A | N/A | N/A | 1400 ksi | **Design Criteria** $\Delta_{total} \leq \dfrac{L}{180} = \dfrac{10 \text{ ft} \left(12 \dfrac{in}{ft}\right)}{180} = 0.67 \text{ in.}$ 0.97 in. < 0.67 in. O.K. $\Delta_{live} \leq \dfrac{L}{240} = \dfrac{10 \text{ ft} \left(12 \dfrac{in}{ft}\right)}{240} = 0.5 \text{ in.}$ 0.27 in. < 0.5 in. O.K.

Case Study 6-3b Select the lightest section based on the section modulus required.

Design Data for (Beam 2) :

Joist or Beam Span = (20 ft)			Tributary Width = (9 ft)		
Loads : D.L. = 15 psf			S.L. = 35 psf		
Species : Douglas Fir Larch			Grade : No 1		

Allowable Stresses :

	Reference Design Values	$(C_F)_{assumed}$	C_D	C_r	Adjusted Value
$F_b' =$	1350 psi	1.0	1.15	1.0	1553 psi
$F_v' =$	170 psi	N/A	1.15	N/A	196 psi
$E' =$	1600 ksi	N/A	N/A	N/A	1600 ksi

From Area Load to Line Load

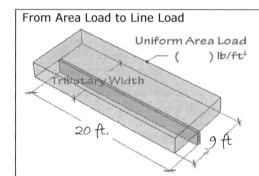

Uniform Area Load ( ) lb/ft²

Tributary Width

20 ft.

9 ft

$$D.L. = 15 \frac{lb}{ft^2} \times 9\,ft = 135 \frac{lb}{ft}$$

$$L.L. = 35 \frac{lb}{ft^2} \times 9\,ft = 315 \frac{lb}{ft}$$

$$T.L. = \qquad\qquad 450\ plf$$

Bending Moment

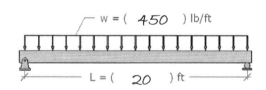

w = (450) lb/ft

L = (20) ft

$$M_{max} = \frac{wL^2}{8} = \frac{\left(450\frac{lb}{ft}\right)(20\,ft)^2}{8} = 22500\ lb-ft$$

Required Section Modulus

$$\left(f_b\right)_{max} = \frac{M_{max}}{S_x} \le F_t' \qquad or, \quad S_x \ge \frac{M_{max}}{F_t'}$$

$$S_x \ge \frac{22500\ lb-ft\ (12\frac{in}{ft})}{1553\frac{lb}{in^2}} = 173.85\ in^3$$

Select Lightest _TRIAL_ Size

	S_x	A	
6 X 16	220.23 in³	85.25 in²	<-- Try the smallest cross-section
8 X 14	227.81 in³	101.25 in²	
10 X 12	209.40 in³	109.25 in²	

Trial Size (6 X 16)

$S_x = 220.23\ in^3$	$I_x = 1706.8\ in^4$	$A = 85.25\ in^2$

Actual	Allowable

Bending Stress

$$(f_b)_{max} = \frac{M_{max}}{S}$$

$$= \frac{22500\ lb-ft\left(12\ \frac{in}{ft}\right)}{220.23\ in^3}$$

$$= 1226\ psi$$

(note: C_F for 6 x 16 must be used.)

		C_F	C_D	C_r	
$F_b' =$	1350	0.972	1.15	1.0	1509 psi

Design Criteria

$$(f_b)_{max} \le F_t'$$

1226 psi < 1509 psi   O.K.

Shear Stress

$$V_{max} = \frac{wL}{2} = \frac{\left(450\ \frac{lb}{ft}\right)(20\ ft)}{2} = 4500\ lb$$

$$(f_v)_{max} = 1.5\ \frac{V_{max}}{A}$$

$$= 1.5\ \frac{4500\ lb}{85.25\ in^2} = 79\ psi$$

		C_F	C_D	C_r	
$F_v' =$	170	N/A	1.15	N/A	196 psi

Design Criteria

$$(f_v)_{max} \le F_v'$$

83 psi < 207 psi   O.K.

Deflection

$$\Delta_{total} = \frac{5\ W_{total}\ L^4\ (12^3)}{384\ EI}$$

$$= \frac{5\ (450)\ (20)^4\ (12^3)}{384(1600000)\ (1706.8)} = 0.59\ in.$$

$$\Delta_{live} = \Delta_{total} \times \frac{W_{live}}{W_{total}} = 0.59\ in. \times \frac{315}{450}$$

$$= 0.41\ in.$$

		C_F	C_D	C_r	
$E' =$	1600	N/A	N/A	N/A	1600 ksi

Design Criteria

$$\Delta_{total} \le \frac{L}{180} = \frac{20\ ft\ (12\ \frac{in}{ft})}{180} = 1.33\ in.$$

0.59 in. < 1.33 in.   O.K.

$$\Delta_{live} \le \frac{L}{240} = \frac{20\ ft\ (12\ \frac{in}{ft})}{240} = 1.0\ in.$$

0.41 in. < 1.0 in.   O.K.

Workshop 6-3a

Design Data
  Dead Load   : 15 psf
  Roof Live Load   : 25 psf

  Roof deflection limit   : $\Delta_T \leq L/180$ and $\Delta_L \leq L/240$.

1. Select the most economical size Joist using Douglas Fir Larch, No 2.

    J1 : _____

2. Select the most economical size for the beam, B3, using Spruce Pine Fur, No2.

    B3 : _____

Workshop 6-3b

Design Data
  Dead Load   : 15 psf
  Floor Live Load   : 60 psf

  Floor deflection limits   : $\Delta_T \leq L/240$ and $\Delta_L \leq L/360$.

1. Select the most economical size Joist using Hem Fir, No 2.

    J4 : _____

2. Select the most economical size for the beam, B3, using Spruce Pine Fur, No1.

    B3 : _____

Workshop 6-3c

Design Data
  Dead Load   : 15 psf
  Snow Load   : 40 psf

  Floor deflection limits   : $\Delta_T \leq L/180$ and $\Delta_L \leq L/240$.

1. Select the most economical size Joist using Spruce Pine Fur, No 2.

    J4 : _____

2. Select the most economical size for the beam, B3, using Douglas-Fir South, No1.

    B5 : _____

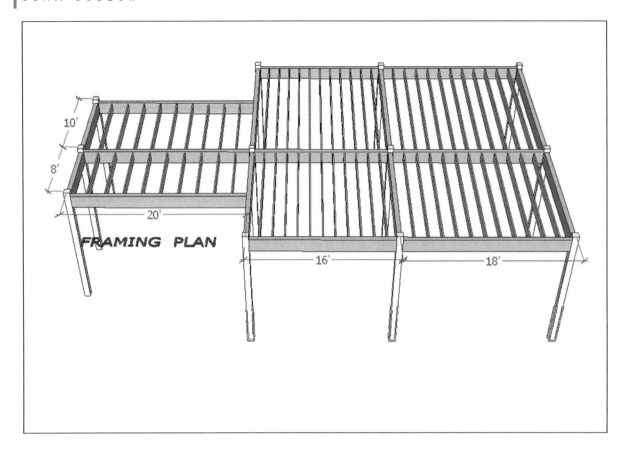

FRAMING PLAN

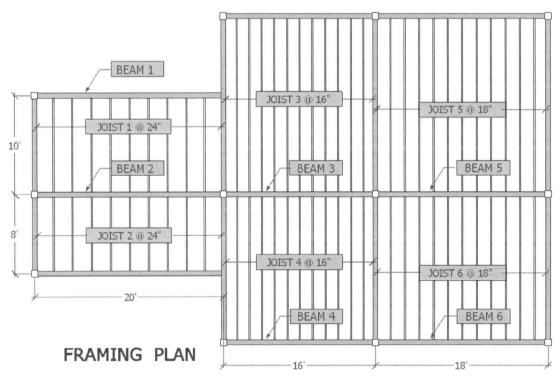

BEAM 1

JOIST 3 @ 16"

JOIST 5 @ 18"

JOIST 1 @ 24"

BEAM 2

BEAM 3

BEAM 5

JOIST 2 @ 24"

JOIST 4 @ 16"

JOIST 6 @ 18"

BEAM 4

BEAM 6

10'

8'

20'

16'

18'

FRAMING PLAN

Workshop 6-3a Select the lightest section based on the section modulus required.

Design Data for (   ) :

Joist or Beam Span = ( )	Tributary Width = ( )
Loads :	
Species	Grade

Allowable Stresses :

	Reference Design Values	$(C_F)_{assumed}$	C_D	C_r	Adjusted Value
$F_b' =$					
$F_v' =$					
$E' =$					

From Area Load to Line Load

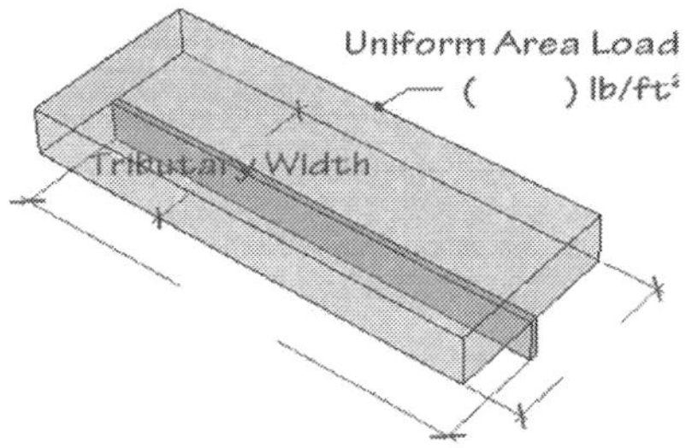

Uniform Area Load (   ) lb/ft^2

Tributary Width

Bending Moment

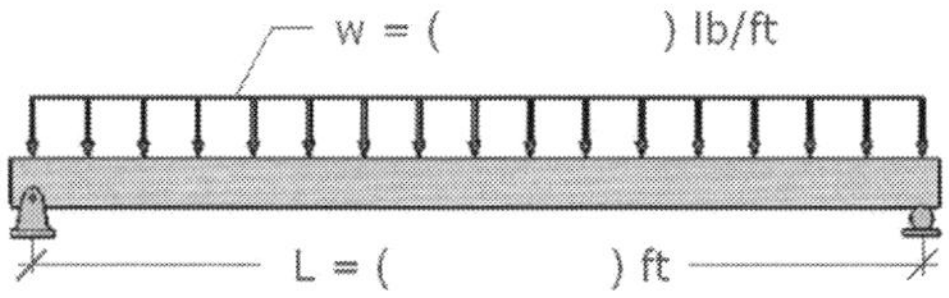

w = (   ) lb/ft

L = (   ) ft

Required Section Modulus

Select Lightest TRIAL Size

Trial Size (   )

$S_x =$	$I_x =$	$A =$

Actual	Allowable
Bending Stress	
Shear Stress	
Deflection $$\Delta_{total} = \frac{5\ W_{total}\ L^4\ (12^3)}{384\ E\ I} =$$ $$\Delta_{live} = \Delta_{total} \times \frac{W_{live}}{W_{total}} =$$	

Select the lightest section based on the section modulus required.

Design Data for (   ) :

Joist or Beam Span = ( )	Tributary Width = ( )
Loads :	
Species	Grade

Allowable Stresses :

	Reference Design values	(C_F) assumed	C_D	C_r	Allowable values
$F_b' =$					
$F_v' =$					
$E' =$					

From Area Load to Line Load

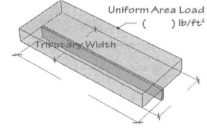

Uniform Area Load
(   ) lb/ft²

Tributary Width

Bending Moment

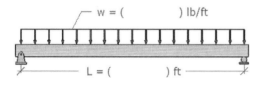

w = (   ) lb/ft

L = (   ) ft

Required Section Modulus

Select Lightest TRIAL Size

Trial Size (   )

$S_x =$	$I_x =$	$A =$

Actual	Allowable
Bending Stress	
Shear Stress	
Deflection $\Delta_{total} = \dfrac{5\,W_{total}\,L^4\,(12^3)}{384\,E\,I} =$ $\Delta_{live} = \Delta_{total} \times \dfrac{W_{live}}{W_{total}} =$	

6.7 Workshop for **Sloped Beam Analysis**

Case Study 6-4a

Given:

The roof dead load is 15 psf and the snow load is 30 psf. The I-joist rafters are spaced at 16 in. on center.

Determine:

The load, shear and bending moment diagrams for a typical I-joist rafter using the sloped roof plane roof method and the horizontal plane method. Hypothetical

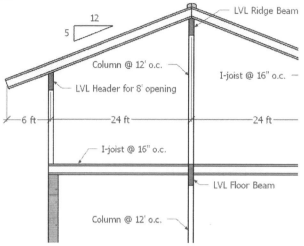

Solution

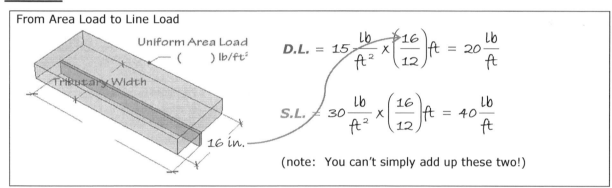

From Area Load to Line Load

Uniform Area Load ( ) lb/ft²

Tributary Width

16 in.

$$D.L. = 15\frac{lb}{ft^2} \times \left(\frac{16}{12}\right)ft = 20\frac{lb}{ft}$$

$$S.L. = 30\frac{lb}{ft^2} \times \left(\frac{16}{12}\right)ft = 40\frac{lb}{ft}$$

(note: You can't simply add up these two!)

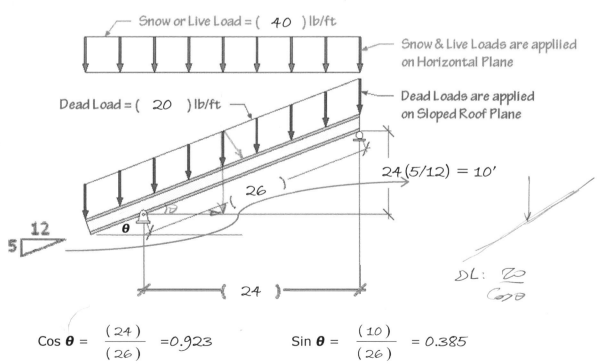

Snow or Live Load = (40) lb/ft

Snow & Live Loads are applied on Horizontal Plane

Dead Load = (20) lb/ft

Dead Loads are applied on Sloped Roof Plane

24 (5/12) = 10'

26

12
5

θ

(24)

DL: $\frac{20}{\cos\theta}$

$$\text{Cos } \theta = \frac{(24)}{(26)} = 0.923 \qquad \text{Sin } \theta = \frac{(10)}{(26)} = 0.385$$

Horizontal Plane Method	Sloped Roof Plane Method

Horizontal Plane Method

S.L. =   (unchanged)   = 40 lb/ft

D.L. = (20 lb/ft) / Cos θ = 21.67 lb/ft

T.L. =   61.67 lb/ft

Hypothetical Beam

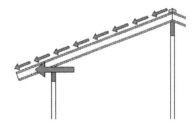

Total Load = (61.67) lb/ft

(6')   (24')

Note:

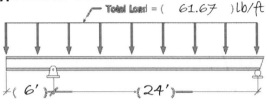

The horizontal tractions do not cause any bending moment of the beam but will increase the axial stresses on the compression zone of the beam cross-sections.

This horizontal thrust on the left pin support is not considered in Horizontal Plane Method but must not be forgotten because this horizontal reaction of the beam in turn pushes the supporting wall outward.

Considering the horizontal components only:

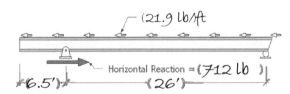

(21.9 lb/ft

Horizontal Reaction = (712 lb)

(6.5')   (26')

Horizontal Reaction from $\Sigma F_x = 0$:

H = 21.9 lb/ft x (6.5' + 26') = 712 lb

Sloped Roof Plane Method

S.L. = (40 lb/ft) × Cos θ = 36.92 lb/ft

D.L. =   (unchanged)   = 20 lb/ft

T.L. =   56.92 lb/ft

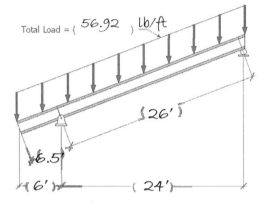

Total Load = (56.92) lb/ft

(26')

6.5'

(6')   (24')

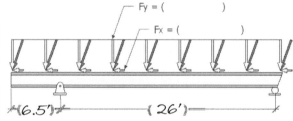

Fy = (   )

Fx = (   )

(6.5')   (26')

Resolving into horz. and vert. components:

F_y = 56.92 lb/ft x Cos θ = 52.5 lb/ft

F_x = 56.92 lb/ft x Sin θ = 21.9 lb/ft

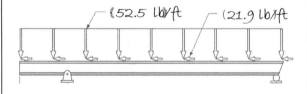

(52.5 lb/ft   (21.9 lb/ft

Considering the vertical components only:

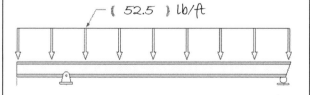

(52.5) lb/ft

❚ B E A M   D E S I G N

Load, Shear Force and Bending Moment Diagrams

Horizontal Plane Method	Sloped Roof Plane Method
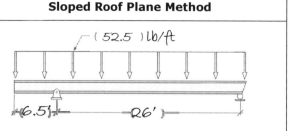 Total Load = (61.67) lb/ft — ‹(6')›‹(24')›	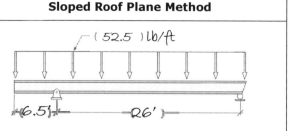 (52.5) lb/ft — ‹6.5›‹ 26' ›
Reactions $$\Sigma M_{(A)} = 0, \quad V_B = \frac{61.67 \frac{lb}{ft} \times 30ft \times 9ft}{24\,ft} = 694\ lb$$ $$\Sigma M_{(B)} = 0, \quad V_A =$$ $$\frac{61.67 \frac{lb}{ft} \times 30ft \times 15ft}{24\,ft} = 1156\ lb$$	Reactions $$\Sigma M_{(A)} = 0, \quad V_B = \frac{52.5 \frac{lb}{ft} \times 32.5' \times 9.75'}{26\,ft} = 640\ lb$$ $$\Sigma M_{(B)} = 0, \quad V_A = \frac{52.5 \frac{lb}{ft} \times 32.5' \times 16.25'}{26\,ft} = 1066\ lb$$
Load Diagrams 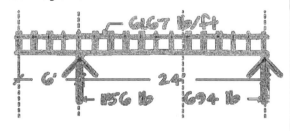	Load Diagrams 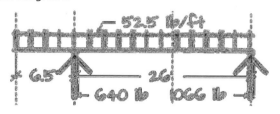
Shear Force Diagram 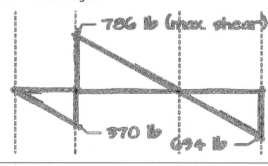	Shear Force Diagram 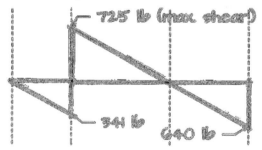
Bending Moment Diagram 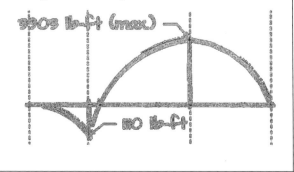	Bending Moment Diagram 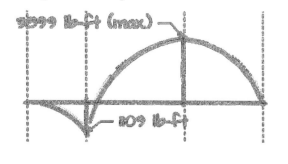

Case Study 6-4b

Given:

The roof dead load is 14 psf and the snow load is 35 psf. The I-joist rafters are spaced at 18 in. on center.

Determine:

The load, shear and bending moment diagrams for a typical I-joist rafter using the sloped roof plane roof method and the horizontal plane method.

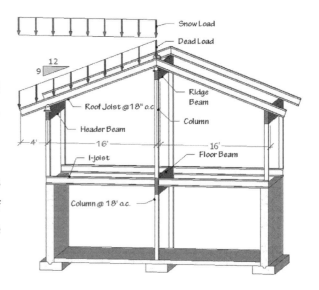

Solution

From Area Load to Line Load

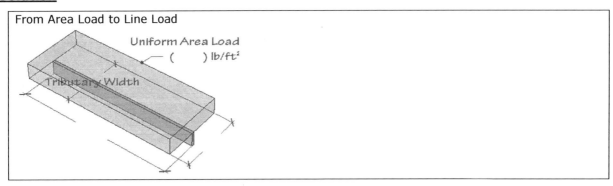

Uniform Area Load
(   ) lb/ft²

Tributary Width

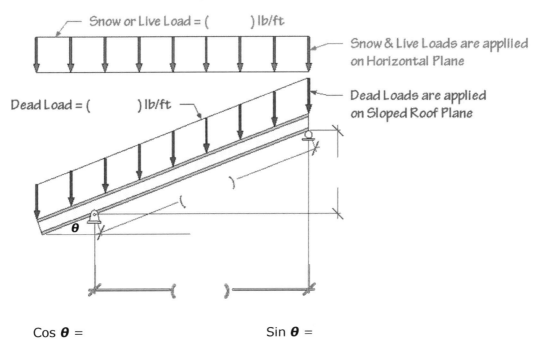

Snow or Live Load = (   ) lb/ft

Snow & Live Loads are applied on Horizontal Plane

Dead Load = (   ) lb/ft

Dead Loads are applied on Sloped Roof Plane

Cos **θ** =   Sin **θ** =

Horizontal Plane Method	Sloped Roof Plane Method
S.L. = (unchanged) = D.L. = ( ) / Cos θ = T.L. =	S.L. = ( ) × Cos θ = D.L. = (unchanged) = T.L. =

Hypothetical Beam

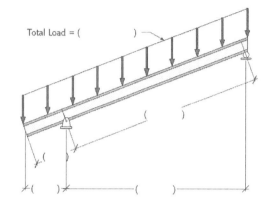

Total Load = (   )

Note:

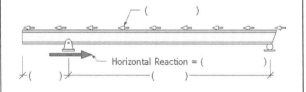

The horizontal tractions do not cause any bending moment of the beam but will increase the axial stresses on the compression zone of the beam cross-sections.

This horizontal thrust on the left pin support is not considered in Horizontal Plane Method but must not be forgotten because this horizontal reaction of the beam in turn pushes the supporting wall outward.

Considering the horizontal components only:

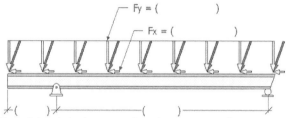

Horizontal Reaction = (   )

Horizontal Reaction from $\Sigma F_x = 0$:

Total Load = (   )

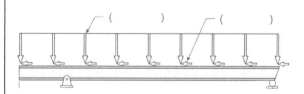

F_y = (   )

F_x = (   )

Resolving into horz. and vert. components:

$F_x =$

$F_y =$

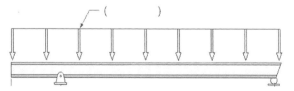

Considering the vertical components only:

Load, Shear Force and Bending Moment Diagrams

Horizontal Plane Method	Sloped Roof Plane Method
Total Load = ( ) ( ) ( )	( ) ( ) ( )
Reactions	Reactions
Load Diagrams	Load Diagrams
Shear Force Diagram	Shear Force Diagram
Bending Moment Diagram	Bending Moment Diagram

Workshop 6-4

Given:

The roof dead load is 16 psf and the snow load is 40 psf. The I-joist rafters are spaced at 24 in. on center.

Determine:

The load, shear and bending moment diagrams for a typical I-joist rafter using the sloped roof plane roof method and the horizontal plane method.

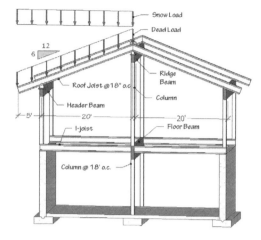

Workshop 6-5

Given:

The roof dead load is 15 psf and the snow load is 30 psf. The I-joist rafters are spaced at 18 in. on center.

Determine:

The load, shear and bending moment diagrams for a typical I-joist rafter using the sloped roof plane roof method and the horizontal plane method.

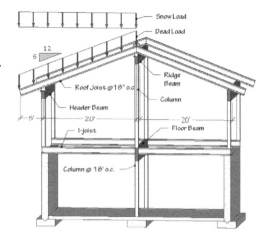

CHAPTER

7

COMPRESSION MEMBER DESIGN

7.1 Concentrically Loaded Member

Compression members are structural elements that are subjected only to concentric compression forces. Loads are applied through the centroid along a longitudinal axis. In this idealized case, the

axial stress f_c is calculated as:

$$f_c = \frac{P}{A}$$

where,

P = compression force
A = cross-sectional area of column

Crushing of Short Column

Note that the ideal state is never realized in practice and some eccentricity of load is inevitable. Thus, compression members fail either by crushing (excessive compressive stress) or by buckling: short members fail by crushing while long slender members fail by buckling. Unless the moment is negligible, the member should be designed as a beam-column. Wood compression member includes columns, wall studs, and compression chords in trusses, shear walls and diaphragm. The design of wood columns or other types of compression members requires consideration of both material failure and stability. The two design factors in wood column design are the compressive strength parallel to the grain and buckling. For the compressive strength consideration, the maximum compressive stress, f_c, induced by loads must not exceed the allowable compressive parallel to the grain, F_c', which is obtained by multiplying the corresponding reference design value by applicable adjustment factors for service conditions. For the instability consideration, buckling failure of columns must be accounted for by including another adjustment factor - column stability factor, C_p.

7.2 Buckling

Buckling is a form of instability. It occurs suddenly with large changes in deformation but little change in loading. For this reason it is a dangerous phenomenon that must be avoided in structural design. Buckling and bending are similar in that they both involve bending moments. In bending, these moments are substantially independent of the resulting deflections, whereas in buckling, the moments and deflections

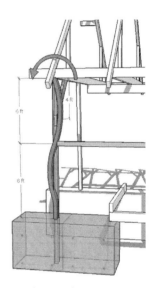

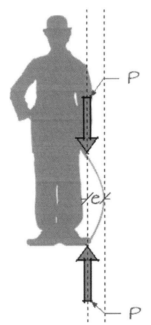

are mutually inter-dependent - so moments, deflections and stresses are not proportional to loads. If the column is not straight (initially crooked), bending moments will develop in the column. Similarly, if the axial load is applied eccentric to the centroid, bending moments (eccentricity moment or P-δ effect), $M = P \times e$, will also develop. If the axial load, P gradually increases, it will ultimately become large enough to cause the member to fail due to instability. This is a geometric consideration, completely divorced from any material strength consideration. If a structural member is prone to buckling then its design must satisfy both strength and buckling safety constraints. Buckling has become more of a problem in recent years since the use of high strength material requires less material for load support - structures and components have become generally more slender and buckle-prone. The concept of buckling may be understood easily as a successive process of bending occurring in a **staple**.

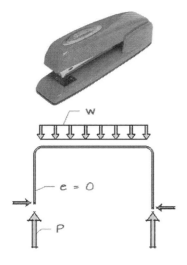

Initially a staple looks like a frame structure without supports at the two ends of the two vertical elements. Both vertical and lateral reactions are applied to the ends of the staple. The lateral reactions, which are induced due to the shape of the grooves, make the two vertical elements of the staple bend and move inwards slightly.

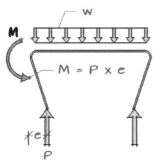

The ends of the staple leave the grooves and no lateral forces act. But, in this position the vertical reactions can continue to bend the two vertical elements due to the sufficient eccentricity and resulting moment.

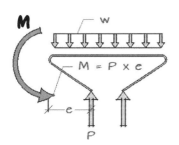

The eccentricity moment make the vertical element bend further producing **additional eccentricity** and resulting **additional moment**. Note that P remains constant.

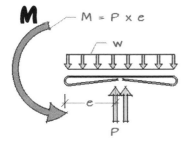

This **viscous cycle of bending** continues until the vertical element **buckle**s. The plastic, permanent deformation is induced.

7.3 **Euler Critical Buckling Load**

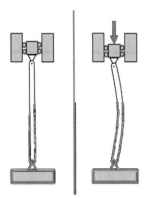

In 1757, mathematician Leonhard Euler derived a formula that gives the maximum axial load that a long, slender, column can carry without buckling. The maximum load, sometimes called the Euler Critical Buckling Load, causes the column to be in a state of unstable equilibrium.

Euler Critical Buckling Load for a *pin-pin connected* column is:

$$P_{cr} = \frac{\pi^2 EI}{l^2}$$

where,   **E** = Modulus of Elasticity

I = Moment of Inertia

l = un-braced length

Euler Critical Buckling Stress is readily obtained by dividing Euler Buckling Load by the cross-section of the compression member.

$$f_{cr} = \frac{P_{cr}}{A} = \frac{\pi^2 EI}{l^2}\frac{1}{A} = \frac{\pi^2 E}{l^2}\frac{I}{A} = \frac{\pi^2 E}{l^2}r^2 = \frac{\pi^2 E}{\left(\dfrac{l}{r}\right)^2}$$

where,   *l* = un-braced length

r = radius of gyration = $\sqrt{\dfrac{I}{A}}$

7.4 **Effective Length Factor**

Euler buckling stress is based on the behavior of a column where both ends are simply supported, or pin-pin connected. The support conditions of compression members in reality may differ from this assumption. In order to be able to apply the formula for Euler Critical Buckling Stress, the concept of an effective length is adopted. The effective length or buckling length of a compression member is defined as the length of a hypothetical pin-pin connected column with the same critical buckling load as the member in question. The equivalent or effective length can be visualized as the distance between two consecutive inflection points of the actual compression member. In practice, an effective length factor, **K** is used which denotes the ratio of the effective length to the unbraced length of the member. Therefore, the Euler critical buckling stress formula becomes:

$$f_{cr} = \frac{\pi^2 E}{\left(\dfrac{Kl}{r}\right)^2}$$

where,

K = effective length factor

Kl = effective length

$\left(\dfrac{Kl}{r}\right)$ = slenderness ratio

7.5 Buckling Loads and End Support Conditions

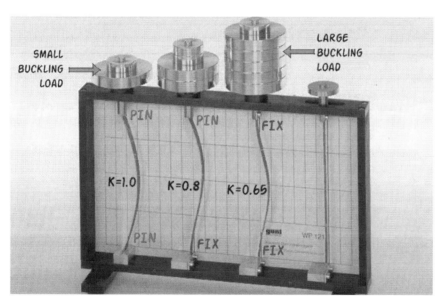

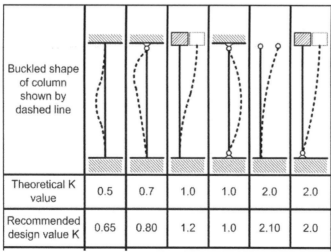

Buckled shape of column shown by dashed line						
Theoretical K value	0.5	0.7	1.0	1.0	2.0	2.0
Recommended design value K	0.65	0.80	1.2	1.0	2.10	2.0

The following demonstrates shows how the boundary conditions of columns affect the corresponding buckling shapes and the buckling capacities of them.

Demonstration of Buckling and Support Conditions

	Put one end of the ruler on the surface of a table and the other end in the palm of one's hand. Press axially on the top end of the ruler and gradually increase the compression force. The straight ruler will suddenly deflect laterally as shown. The deformation becomes larger with further application of the compressive force. This simulates the buckling of a column with two **pinned ends**.
	Now hold the two ends of the ruler tightly to prevent any **rotational** and lateral movements of the two ends of the ruler. Then gradually press axially on the ruler using the fingers until the ruler deforms sideways as shown. This demonstrates a **different buckling shape** for a column with two **ends fixed**. One can clearly feel that a **larger force** is needed in this demonstration than in the previous demonstration.
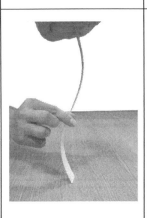	If one intermediate **lateral support** is provided, so that the ruler cannot move sideways at this point and a **different buckling shape** is created. Even larger compressive force will be required to make the ruler buckle in the shape shown.

7.6 Wood Column Design

The column design criterion for Allowable Stress Design (ASD) is that the stress in a member must be less than an allowable stress which is equal to the yield stress divided by a factor of safety.

$$f_c = \frac{P}{A} \leq F_c'$$

f_c = actual compressive stress parallel to the grain

F_c' = allowable compressive stress parallel to the grain

$F_c' = F_c^* \times C_p = [F_c \, (C_F)(C_D)(C_M)(C_t)(C_i)] \times C_p$

F_c = compressive strength parallel to the grain

C_F = size factor

C_D = load duration factor

C_M = wet service factor

C_t = temperature factor

C_p = column stability factor

For wood columns, the ratio of the column length to its width is just as important as the slenderness ratio is for steel columns. In wood columns, the slenderness ratio is defined as the laterally unsupported length in inches divided by the depth (dimension in the same direction of the plane of buckling) of the column. Wood columns are restricted to a maximum slenderness ratio of $Kl/d \leq 50$. This slenderness ratio is analogous to the limiting slenderness ratio of $Kl/r \leq 200$ used for steel columns.

Procedure for Analysis

1. Calculate the controlling slenderness ratio, $\left(\dfrac{Kl}{d}\right) \leq 50$

2. Calculate $F_c^* \times C_p = [F_c \, (C_F)(C_D)(C_M)(C_t)(C_i)] \times C_p$

3. Calculate $F_{cE} = \dfrac{0.822\, E'_{min}}{\left(\dfrac{Kl}{d}\right)^2}$

   E'_{min} = adjusted minimum modulus of elasticity

4. Calculate $\dfrac{F_{cE}}{F_c^*}$

5. Get C_p from Appendix, Table B-6

6. Calculate the allowable compressive stress, $F_c' = F_c^* \times C_p$

7. Calculate the allowable compressive load, $P' = F_c' \times A$

7.7 Workshop for **Wood Column Design**

Case Study 7-1 Determine the allowable compressive load that the 6 x 8 Hem-Fir No. 2 grade column can support. Consider DL+RLL.

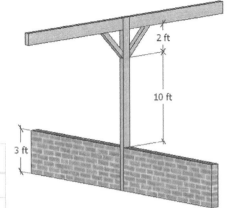

Section Properties

thk = 5.5 in	width = 7.5 in	A = 41.25 in²

Adjusted Design Values

	Ref. Value	C_D	C_F	C_t	C_M	C_i	Adjusted Values
F_c^*=	575	1.25	1.0	1.0	1.0	1.0	718.8 psi
E'_{min}=	400000	n/a	1.0	1.0	1.0	1.0	400000 psi

Slenderness Ratio

x-axis buckling		K_x = 1.0	Note that the wall doesn't provide lateral supports to the column. Why? L_x = 13 ft	$\left(\dfrac{Kl}{d}\right)_x = \dfrac{1.0\,(13\,ft \times 12)}{7.5\,in} = 20.8$
y-axis buckling		K_y = 0.8 L_y = 12 ft	**Larger** slenderness ratio **controls** !	$\left(\dfrac{Kl}{d}\right)_y = \dfrac{0.8\,(12\,ft \times 12)}{5.5\,in} = 20.95$

Design Parameters

$$F_{cE} = \frac{0.822\,E'_{min}}{\left(\dfrac{Kl}{d}\right)^2} = \frac{0.822\,(400000)}{(20.95)^2} = 749.5\ psi$$

$$\frac{F_{cE}}{F_c^*} = \frac{749.35\ psi}{718.8\ psi} = 1.0427$$

Column Stability Factor, C_p (c = 0.8 for glulam)

$$C_P = \frac{1 + \dfrac{F_{cE}}{F_c^*}}{2c} - \sqrt{\left(\frac{1 + \dfrac{F_{cE}}{F_c^*}}{2c}\right)^2 - \frac{\dfrac{F_{cE}}{F_c^*}}{c}}$$

or, from Appendix B, Table B-6 with linear interpolation C_p = 0.7051

Allowable Compressive Stress $F_c' = F_c^* \times (C_P)$ = 718.8 psi × 0.70526 = 506.9 psi

Allowable Compressive Load $P = F_c'\,(A)$ = 506.9 psi × 41.25 in² = 20909 lb = 20.9 k

Workshop 7-1a

Determine the allowable compressive load that the Douglas Fir Larch No. 2 grade column (6x8) can support. Consider DL+SL.

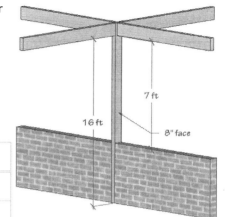

Section Properties

$b =$	$d =$	$A =$

Adjusted Design Values

	Ref. Values	C_D	C_F	C_t	C_M	C_i	Adjusted Values
$F_c^* =$							
$E'_{min} =$							

Slenderness Ratio

x-axis buckling			
	$K_x =$	$L_x =$	$\left(\dfrac{Kl}{d}\right)_x =$
y-axis buckling			
	$K_y =$	$L_y =$	$\left(\dfrac{Kl}{d}\right)_y =$

Design Parameters

$$F_{cE} = \frac{0.822\, E'_{min}}{\left(\dfrac{Kl}{d}\right)^2} = \qquad\qquad \frac{F_{cE}}{F_c^*} =$$

Column Stability Factor, C_P

$$C_P = \frac{1 + \dfrac{F_{cE}}{F_c^*}}{2c} - \sqrt{\left(\frac{1 + \dfrac{F_{cE}}{F_c^*}}{2c}\right)^2 - \frac{\dfrac{F_{cE}}{F_c^*}}{c}}$$

or, *from Appendix B, Table B-6 with linear interpolation* $C_P =$

Allowable Compressive Stress $F_c' = F_c^* \times (C_P) =$

Allowable Compressive Load $P = F_c'(A) =$

Workshop 7-1b

Determine the allowable compressive load that the 2x4 Douglas Fir South No. 2 grade stud wall (7 studs) can support. Consider DL+FL.

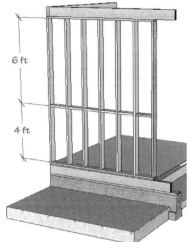

6 ft

4 ft

Section Properties

$b =$	$d =$	$A =$

Adjusted Design Values

	Ref. Values	C_D	C_F	C_t	C_M	C_i	Adjusted Values
$F_c^* =$							
$E'_{min} =$							

Slenderness Ratio

x-axis buckling			
	$K_x =$	$L_x =$	$\left(\dfrac{Kl}{d}\right)_x =$
y-axis buckling			
	$K_y =$	$L_y =$	$\left(\dfrac{Kl}{d}\right)_y =$

Design Parameters

$$F_{cE} = \frac{0.822\, E'_{min}}{\left(\dfrac{Kl}{d}\right)^2} = \qquad\qquad \frac{F_{cE}}{F_c^*} =$$

Column Stability Factor, C_p

$$C_p = \frac{1+\dfrac{F_{cE}}{F_c^*}}{2c} - \sqrt{\left(\frac{1+\dfrac{F_{cE}}{F_c^*}}{2c}\right)^2 - \frac{\dfrac{F_{cE}}{F_c^*}}{c}} \qquad \text{or, from Appendix B, Table B-6 with linear interpolation} \quad C_p =$$

Allowable Compressive Stress $F_c' = F_c^* \times (C_P) =$

Allowable Compressive Load $P = F_c'(A) =$

Workshop 7-1c

Determine the allowable compressive load that the Spruce Pine Fir Select Structural grade corner column (4x6) can support. Consider DL+WL.

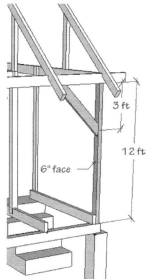

3 ft

12 ft

6" face

Section Properties

$b =$	$d =$	$A =$

Adjusted Design Values

	Ref. Values	C_D	C_F	C_t	C_M	C_i	Adjusted Values
$F_c^* =$							
$E'_{min} =$							

Slenderness Ratio

x-axis buckling			
	$K_x =$	$L_x =$	$\left(\dfrac{Kl}{d}\right)_x =$
y-axis buckling			
	$K_y =$	$L_y =$	$\left(\dfrac{Kl}{d}\right)_y =$

Design Parameters

$$F_{cE} = \frac{0.822\, E'_{min}}{\left(\dfrac{Kl}{d}\right)^2} = \qquad\qquad \frac{F_{cE}}{F_c^*} =$$

Column Stability Factor, C_p

$$C_P = \frac{1 + \dfrac{F_{cE}}{F_c^*}}{2c} - \sqrt{\left(\frac{1 + \dfrac{F_{cE}}{F_c^*}}{2c}\right)^2 - \frac{\dfrac{F_{cE}}{F_c^*}}{c}}$$

or, *from Appendix B, Table B-6 with linear interpolation* $C_p =$

Allowable Compressive Stress $F_c' = F_c^* \times (C_P) =$

Allowable Compressive Load $P = F_c'\,(A) =$

Workshop 7-1d

Determine the allowable compressive load that the Douglas Fir Larch No. 1 grade corner column (6x8) can support. Consider DL+RLL.

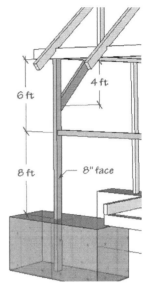

6 ft

4 ft

8 ft

8" face

Section Properties

$b =$	$d =$	$A =$

Adjusted Design Values

	Ref. Values	C_D	C_F	C_t	C_M	C_i	Adjusted Values
$F_c* =$							
$E'_{min} =$							

Slenderness Ratio

x-axis buckling			
	$K_x =$	$L_x =$	$\left(\dfrac{Kl}{d}\right)_x =$
y-axis buckling			
	$K_y =$	$L_y =$	$\left(\dfrac{Kl}{d}\right)_y =$

Design Parameters

$$F_{cE} = \frac{0.822\, E'_{min}}{\left(\dfrac{Kl}{d}\right)^2} =$$

$$\frac{F_{cE}}{F_c^*} =$$

Column Stability Factor, C_p

$$C_p = \frac{1+\dfrac{F_{cE}}{F_c^*}}{2c} - \sqrt{\left(\frac{1+\dfrac{F_{cE}}{F_c^*}}{2c}\right)^2 - \frac{\dfrac{F_{cE}}{F_c^*}}{c}}$$

or, *from Appendix B, Table B-6 with linear interpolation* $C_p =$

Allowable Compressive Stress $F_c' = F_c* \times (C_p) =$

Allowable Compressive Load $P = F_c'(A) =$

COMBINED BENDING AND AXIAL FORCE

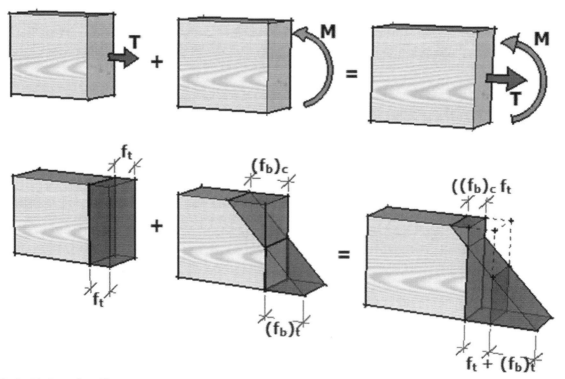

8.1 Introduction

When bending moment take place with an axial tension force, the effects of combined stresses must be considered. In the example of the bottom chords of a truss subjected to gravity loads, the axial tensile stresses and bending tensile stresses add in the bottom extreme fiber. However, the axial tensile stresses and bending compressive stresses counter in the top extreme fiber. Thus, depending on the relative magnitudes of the stresses, the resultant stress on the face of the cross section can be either tension or compression. This type of combined loading can be governed by either a combined tension criterion or a net compression criterion.

8.2 Combined Axial Tension and Bending Tension

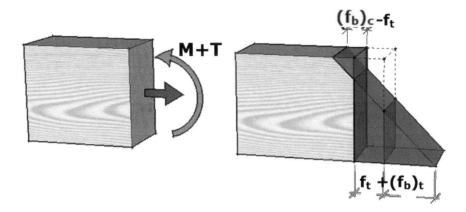

Combined tensile and bending stresses are analyzed using a linear interaction equation as shown below. For allowable stress design (ASD) procedure:

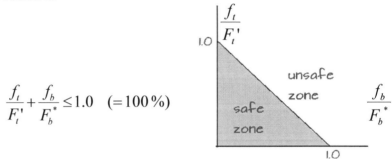

$$\frac{f_t}{F_t'} + \frac{f_b}{F_b^*} \leq 1.0 \quad (= 100\%)$$

interaction curve

where

f_t = actual working tensile stress parallel to the grain

F_t' = adjusted tensile design value

f_b = actual working bending stress

F_b^* = adjusted bending design value without the adjustment for lateral stability (since the actual bending stress is the bending tensile stress).

$F_b^* = F_b \times C_D \times C_M \times C_t \times C_F \times C_{fu} \times C_i \times C_r$ for sawn lumber

$\quad\;\; = F_b \times C_D \times C_M \times C_t \times C_V \times C_{fu} \times C_c$ for glulam

The term, $\dfrac{f_t}{F_t'}$ in the equation is known as a *stress ratio* and measures the effects of tensile stress. This term in the interaction equation may be interpreted as 'how many percent of the member's total capacity is used to resist axial force' while $\dfrac{f_b}{F_b^*}$ measures the effects of bending stress.

8.3 **Combined Bending and Compression**

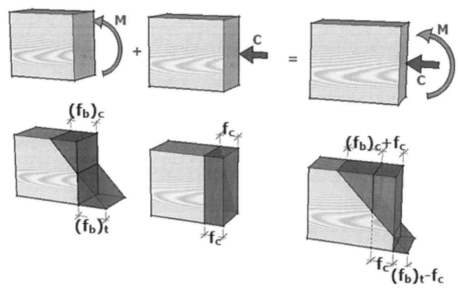

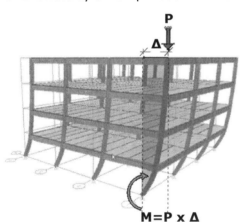

Structural members stressed simultaneously in bending and compression are known as beam-columns. A straight-line interaction formula was used to account for the interaction between bending moment and compression force. However, 2005 editions of the NDS use a modified version of the interaction equation to account for the **P-Δ** effect by an amplification factor in the analysis of beam-columns.

A typical architectural example of the **P-Δ** effect is a laterally displaced multi-story building structure with significant gravity loads acting on it. When a building in an earthquake horizontally drifs far enough so that the gravity loads start to act at the new positions offset from the initial positions pulling the building laterally further (even bigger **Δ**). The additional builidng drift is caused by the '*destabilizing*' moment which is calculated M = P x Δ. Note that the gravity loads actiong on the builind, P is constant, while the building dirfit, Δ keeps increasing which in turn creates bigger destabilizing moment, M. The detimental process may be continued until the builidng is collapsed due to too large story drifts unless it has enough lateral stiffness and strength. The **P-Δ** effect not only magnifies story drift but also affects certain structural behaviors while reducing deformation capacity (stiffness) of a building. An accurate structural analysis must account for the **P-Δ** effects whenever the 'small displacement' assumption is not valid any longer.

The general interaction formula for the allowable stress design (ASD) procedure with bending only about the x-axis follows:

$$\left(\frac{f_t}{F_t{}'}\right) + \left(\frac{1}{1-f_c/F_{cEx}}\right)\left(\frac{f_b}{F_b{}^*}\right) \leq 1.0$$

AMPLIFICATION FACTOR
FOR COMPRESSION

where

f_c = actual compressive stress = **P/A**

$F_c{}'$ = adjusted compressive design value

= F_c (C_D) (C_M) (C_t) (C_F) (C_i) (C_P) for sawn lumber

= F_c (C_D) (C_M) (C_t) (C_P) for glulam

f_{bx} = actual working bending stress= M_x/S_x

$F_{bx}{}'$ = adjusted bending design value for sawn lumber about the x-axis considering the effects of lateral torsional buckling

= F_b (C_D) (C_M) (C_t) (C_L) (C_F) (C_{fu}) (C_i) (C_r) for sawn lumber

$F_{bx}{}'$ = adjusted bending design value for glulam about the x-axis considering the effects of lateral torsional buckling (smaller of the following bending stress values applies)

= F_{bx} (C_D) (C_M) (C_t) (C_L) (C_{fu}) (C_c) or

= F_{bx} (C_D) (C_M) (C_t) (C_V) (C_{fu}) (C_c)

F_{cEx} = Euler based elastic buckling stress, based on the slenderness ratio for the x-axis $(l_e/d)_x$

$$= \frac{0.822 \, E_{x-min}}{F_t{}'}$$

Note that the interaction equation is not a linear combination of the effects of bending and axial force as was the case of combined bending and tension. This is because bending and tension don't have a detrimental 'amplifying' interaction. As a matter of fact, a tension force has a beneficial 'stiffening' effect on the behavior of tension members. On the contrary, a compression force on a column has a detrimental 'weakening' effect on the column stiffness. To take this 'weakening' effect or the **P-Δ** effects, an amplification factor is introduced to the interaction formula.

Thus, the $\dfrac{1}{\left(1-f_c/F_{cEx}\right)}$ in the interaction formula may be thought of as amplification factor.

8.4 Workshop for **Combined Bending and Tension**

Case Study 8-1 Determine the design adequacy of the lower chord member.

1. The loads from the roof system are applied to the top chords as a uniform load.
2. The dead weight of the ceiling hanging from the bottom chords of the truss is represented by a uniformly distributed load.
3. No. 1 & Better grade Hem-Fir 2x8 is used for the bottom chords.
4. Load combination D.L.+S.L. governs the design of the bottom chords.
5. Connections are made by a single row of ¾" bolts.
6. The trusses are laterally braced and thus lateral buckling needs not be considered.

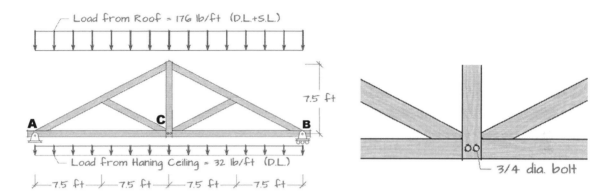

Line loads to Joint loads

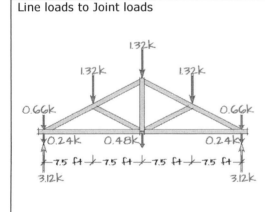

Top Chords

Edge joints :   176 lb/ft x 3.75 ft = 0.66 k

Interior joints : 176 lb/ft x 7.5 ft = 1.32 k

Bottom Chords

Edge joints :   32 lb/ft x 7.5 ft = 0.24 k

middle joints : 32 lb/ft x 15 ft = 0.48 k

Truss Analysis (Method of Joints)

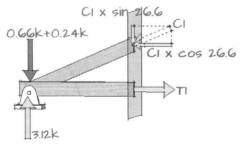

$\Sigma F_y = 0$ gives;

$\quad +3.12k - (0.66k + 0.24k) - C1(\sin 26.6) = 0$

$\quad C1 = 2.22k/(\sin 26.6) = 4.96 k$

$\Sigma F_x = 0$ gives;

$\quad T1 - (4.96k)(\cos 26.6) = 0$

$\quad T1 = 4.44k = 4440 lb$

1. Tension at Net Cross-sectional Area

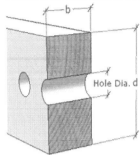

$$A_n = A_g - A_h = b\,d - b\,(D_h) = b\,(d - D_h)$$

$$D_h = \text{bolt diameter} + \frac{1}{16} = \frac{3}{4} + \frac{1}{16} = 0.8125$$

$$A_n = 1.5\,(7.25 - 0.8125) = 9.66 \text{ in}^2$$

Hole Dia. d

Actual Tensile Stress	$f_t = \dfrac{P}{A_n} = \dfrac{4440 \text{ lb}}{9.66 \text{ in}^2} = 460 \text{ psi}$
Allowable Tensile Stress	$F_t' = F_t\,(C_F)\,(C_D)\,(C_M)\,(C_t)\,(C_i)$ $= 725\,(1.2)\,(1.15)\,(1.0)\,(1.0)\,(1.0) = 1000 \text{ psi}$
Decision **1**	$460 \text{ psi} < 1000 \text{ psi}$ O.K. (or, $\dfrac{460 \text{ psi}}{1000 \text{ psi}} = 0.46 < 1.0$ O.K.)

2. Bending in Member AC (or BC)

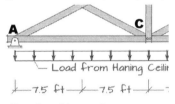

Load from Haning Ceilir

7.5 ft 7.5 ft

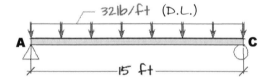

32 lb/ft (D.L.)

A   C

15 ft

Maximum Bending Moment

$$M_{max} = \frac{w\,L^2}{8} = \frac{\left(32\,\dfrac{lb}{ft}\right)(15 \text{ ft})^2}{8} = 900 \text{ lb} - ft$$

Maximum Actual Bending Stress

$$\left(f_b\right)_{max} = \frac{M_{max}}{S_x} = \frac{900 \text{ lb} - ft\,(12)}{13.14 \text{ in}^3} = 822\,\frac{lb}{in^2}$$

Allowable Bending Stress (*Note : C_D is for dead load only.*)

$$F_b{}^{**} = F_b\,(C_F)\,(C_D)\,(C_M)\,(C_t)\,(C_r)\,(C_i)$$

$$= 1100\,(1.2)\,(0.9)\,(1.0)\,(1.0)\,(1.0)\,(1.0) = 1188 \text{ psi}$$

Decision **2**	$822 \text{ psi} < 1188 \text{ psi}$ O.K. (or, $\dfrac{822 \text{ psi}}{1188 \text{ psi}} = 0.69 < 1.0$ O.K.)

3. Combined Bending and Tension in Member AC (or BC)

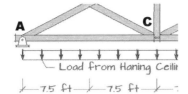

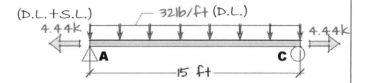

Allowable Bending Stress (*Note : C_D is for the governing load combination D.L. + S.L.*)

$$F_b^* = F_b' = F_b \, (C_F) \, (C_D) \, (C_M) \, (C_t) \, (C_r) \, (C_i)$$

$$= 1100 \, (1.2) \, (1.15) \, (1.0) \, (1.0) \, (1.0) \, (1.0) = 1518 \text{ psi}$$

Actual Tensile Stress (*Note : There is no bolt hole in the bottom chord, AC. Thus the entire cross-section is effective.*)

$$f_t = \frac{P}{A_g} = \frac{4440 \text{ lb}}{(1.5 \times 7.25) \text{ in}^2} = 408 \text{ psi}$$

Allowable Tensile Stress

$$F_t' = F_t \, (C_F) \, (C_D) \, (C_M) \, (C_t) \, (C_i)$$

$$= 725 \, (1.2) \, (1.15) \, (1.0) \, (1.0) \, (1.0) = 1000 \text{ psi}$$

Decision **3** - Interaction Equation

$$\frac{f_t}{F_t'} + \frac{f_b}{F_b'} = \frac{408}{1000} + \frac{822}{1518} = 0.95 \leq 1.0 \ \text{ O.K.}$$

Case Study 8-2 - Combined Bending and Tension

Determine the design adequacy of the lower chord member using ASD procedures.

1. The loads from the roof system are applied to the top chords as a uniform load.
2. The dead weight of the ceiling hanging from the bottom chords of the truss is represented by a uniformly distributed load.
3. No. 1 grade Douglas Fir Larch 2x3 is used for the bottom chords.
4. Load combination D.L.+R.L.L. governs the design of the bottom chords.
5. Connections are made by a single row of ¾" bolts.
6. The trusses are laterally braced and thus lateral buckling needs not be considered.

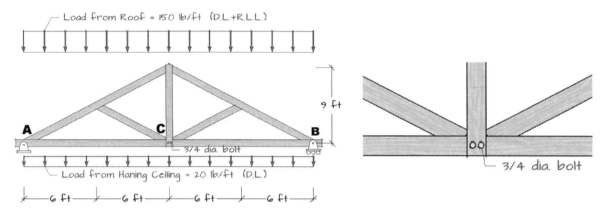

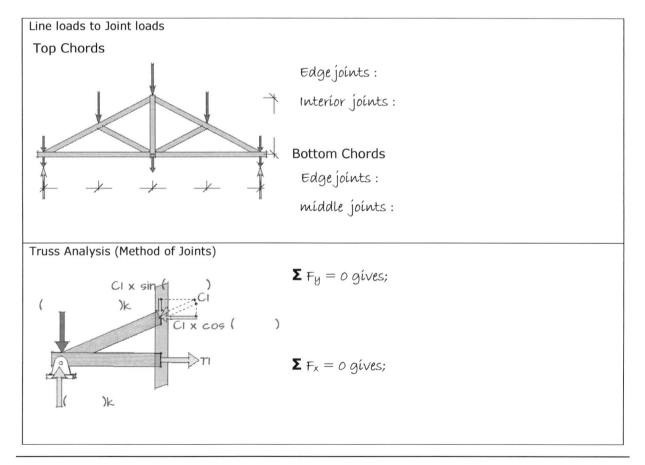

Line loads to Joint loads
Top Chords

Edge joints :

Interior joints :

Bottom Chords

Edge joints :

middle joints :

Truss Analysis (Method of Joints)

$\Sigma\, F_y = 0$ gives;

$\Sigma\, F_x = 0$ gives;

1. Tension at Net Cross-sectional Area

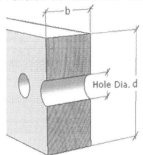

$$A_n = A_g - A_h = b\,d - b\,(D_h) = b\,(d - D_h)$$

$$D_h = \text{bolt diameter} + \frac{1}{16} =$$

$$A_n =$$

Actual Tensile Stress

$$f_t = \frac{P}{A_n} =$$

Allowable Tensile Stress

$$F_t' = F_t\,(C_F)\,(C_D)\,(C_M)\,(C_t)\,(C_i)$$

$$=$$

Decision **1**

---

2. Bending in Member AC (or BC)

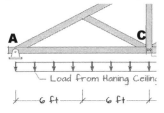

Load from Haning Ceilin

6 ft ⟋ 6 ft

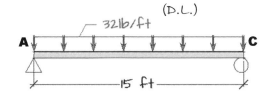

(D.L.)

32 lb/ft

A   C

15 ft

Maximum Bending Moment

$$M_{max} = \frac{wL^2}{8} =$$

Maximum Actual Bending Stress

$$\left(f_b\right)_{max} = \frac{M_{max}}{S_x} =$$

Allowable Bending Stress *(Note : C_D is for dead load only.)*

$$F_b^{**} = F_b\,(C_F)\,(C_D)\,(C_M)\,(C_t)\,(C_r)\,(C_i)$$

$$=$$

Decision **2**

3. Combined Bending and Tension in Member AC (or BC)

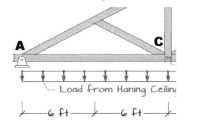

Load from Haning Ceilina

6 ft   6 ft

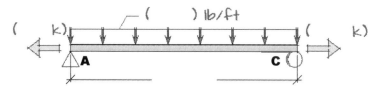

(   ) lb/ft

(k)   (k)

A   C

Allowable Bending Stress *(Note : C_D is for the governing load combination D.L. + S.L.)*

$$F_b^* = F_b' = F_b \ (C_F) \ (C_D) \ (C_M) \ (C_t) \ (C_r) \ (C_i)$$

$$=$$

Actual Tensile Stress *(Note : There is no bolt hole in the bottom chord, AC. Thus the entire cross-section is effective.)*

$$f_t = \frac{P}{A_g} =$$

Allowable Tensile Stress $F_t' = F_t \ (C_F) \ (C_D) \ (C_M) \ (C_t) \ (C_i)$

$$=$$

Decision 3 - Interaction Equation

$$\frac{f_t}{F_t'} + \frac{f_b}{F_b^*} =$$

Workshop 8-1a - Combined Bending and Tension

Determine the design adequacy of the lower chord member using ASD procedures.
1. The loads from the roof system are applied to the top chords as a uniform load.
2. The dead weight of the ceiling hanging from the bottom chords of the truss is represented by a uniformly distributed load.
3. No. 1 grade Douglas Fir South 2x6 is used for the bottom chords.
4. Load combination D.L.+S.L. governs the design of the bottom chords.
5. Connections are made by a single row of ¾" bolts.
6. The trusses are laterally braced and thus lateral buckling needs not be considered.

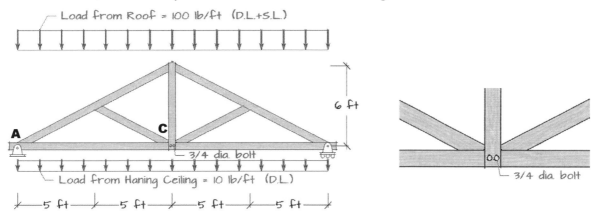

Line loads to Joint loads			

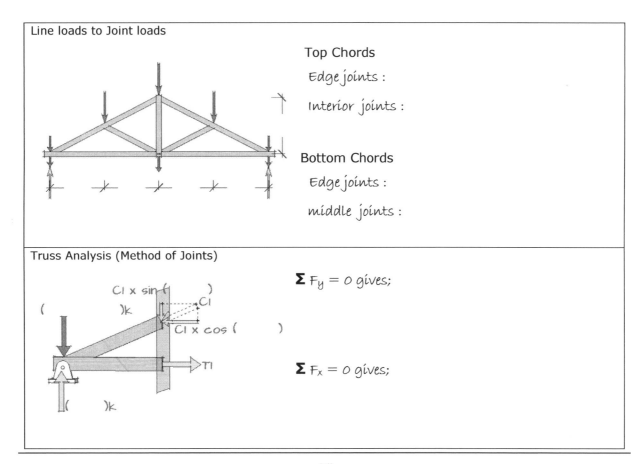

Top Chords

    Edge joints :

    Interior joints :

Bottom Chords

    Edge joints :

    middle joints :

Truss Analysis (Method of Joints)

$\Sigma\, F_y = 0$ gives;

$\Sigma\, F_x = 0$ gives;

1. Tension at Net Cross-sectional Area

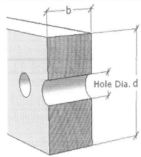

$$A_n = A_g - A_h = b\,d - b\,(D_h) = b\,(d - D_h)$$

$$D_h = bolt\ diameter + \frac{1}{16} =$$

$$A_n =$$

Actual Tensile Stress

$$f_t = \frac{P}{A_n} =$$

Allowable Tensile Stress

$$F_t{}' = F_t\ (C_F)\ (C_D)\ (C_M)\ (C_t)\ (C_i)$$

$$=$$

Decision **1**

2. Bending in Member AC (or BC)

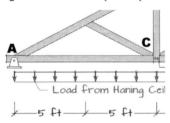

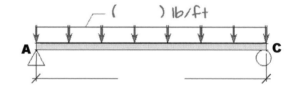

Maximum Bending Moment

$$M_{max} = \frac{w\,L^2}{8} =$$

Maximum Actual Bending Stress

$$\left(f_b\right)_{max} = \frac{M_{max}}{S_x} =$$

Allowable Bending Stress (*Note : C_D is for dead load only.*)

$$F_b{}^{**} = F_b\ (C_F)\ (C_D)\ (C_M)\ (C_t)\ (C_r)\ (C_i)$$

$$=$$

Decision **2**

3. Combined Bending and Tension in Member AC (or BC)

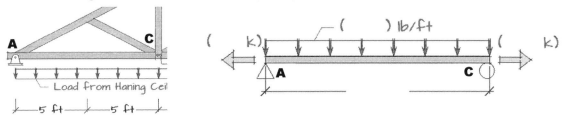

Load from Haning Ceil

5 ft   5 ft

Allowable Bending Stress (*Note : C_D is for the governing load combination, D.L. + S.L.*)

$$F_b{}^* = F_b' = F_b \ (C_F) \ (C_D) \ (C_M) \ (C_t) \ (C_r) \ (C_i)$$

$$=$$

Actual Tensile Stress (*Note : There is no bolt hole in the bottom chord, AC. Thus the entire cross-section is effective.*)

$$f_t = \frac{P}{A_g} =$$

Allowable Tensile Stress $F_t' = F_t \ (C_F) \ (C_D) \ (C_M) \ (C_t) \ (C_i)$

$$=$$

Decision **3** - Interaction Equation

$$\frac{f_t}{F_t{}'} + \frac{f_b}{F_b{}^*} =$$

Workshop 8-1b - Combined Bending and Tension

Determine the design adequacy of the lower chord member using ASD procedures.

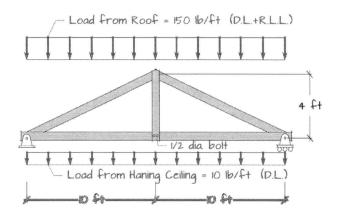

1. The loads from the roof system are applied to the top chords as a uniform load.
2. The dead weight of the ceiling hanging from the bottom chords of the truss is represented by a uniformly distributed load.
3. No. 1 grade Douglas Fir South 2x4 is used for the bottom chords.
4. Load combination D.L.+R.L.L. governs the design of the bottom chords.
5. Connections are made by a single row of 1/2" bolts.
6. The trusses are laterally braced and thus lateral buckling needs not be considered.

Workshop 8-1c - Combined Bending and Tension

Determine the design adequacy of the lower chord member using ASD procedures.

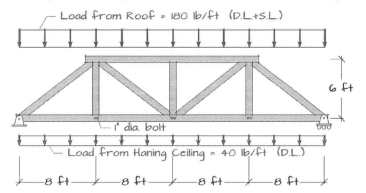

1. The loads from the roof system are applied to the top chords as a uniform load.
2. The dead weight of the ceiling hanging from the bottom chords of the truss is represented by a uniformly distributed load.
3. Select Structural grade South Hemlock 2x8 is used for the bottom chords.
4. Load combination D.L.+S.L. governs the design of the bottom chords.
5. Connections are made by a single row of ¾" bolts.
6. The trusses are laterally braced and thus lateral buckling needs not be considered.

CHAPTER

9

ENGINEERED WOOD PRODUCTS

9.1 Introduction

Engineered wood includes a range of derivative wood products which are manufactured by bonding together wood strands, veneers or laminations with adhesives, to produce a larger and integral composite unit with enhanced structural properties. Engineered wood products (EWP) are engineered to precise design specifications which are tested to meet national or international standards. Typically, engineered wood products are made from the same hardwoods and softwoods used to manufacture sawn lumber. Sawmill scraps and other wood waste can be used for engineered wood composed of wood particles or fibers, but whole logs are usually used for veneers, such as plywood, MDF or Particle board. Some engineered wood products, like oriented strand board (OSB), can use trees from the poplar family, a common but non-structural species. Engineered wood products are used in a variety of ways, often in applications similar to solid wood products.

9.2 Advantages of EWP

Engineered wood products may be preferred over solid wood in some applications due to certain comparative advantages:

1. Higher and uniform structural quality members can be obtained by removing or dispersing growth defects and the natural strength and stiffness characteristics of wood can be maximized.

2. The products are more durable and dimensionally stable than typical wood building materials.

3. Glued laminated timber (glulam) has greater strength and stiffness than comparable dimensional lumber and higher strength-to-weight ratios than steel.

4. Engineered lumber can speed installation time and reduce labor since they are lighter and can be spaced further.

5. Some engineered wood products offer more architectural design options without sacrificing structural requirements.

6. Engineered wood products make more efficient use of wood. They can be made from small pieces of wood, wood that has defects or underutilized species.

7. Engineered wood products are versatile and available in a wide variety of thicknesses, sizes, grades, and exposure durability classifications, making the products ideal for use in unlimited construction, industrial and home project application.

8. Because engineered wood is man-made, it can be designed to meet application-specific performance requirements.

9.3 Disadvantages of EWP

Engineered wood products also have some disadvantages:

1. Some engineered wood products, such as those specified for interior use, may be weaker and more prone to humidity-induced warping than equivalent solid woods. Most particle and fiber-based boards are not appropriate for outdoor use because they readily soak up water.

2. The adhesives used in some products may be toxic. Cutting and otherwise working with some products can expose workers to toxic compounds.

3. Some products may burn more quickly than solid lumber.

4. They require more primary energy for their manufacture than solid lumber.

9.4 **Glued Laminated Timber**

Sawn lumber is manufactured in a large number of sizes and grades and is used for a wide variety of structural members. However, the span becomes long or when the loads become large; the use of sawn lumber may become impractical. Because of size limitation of sawn lumber or other structural and architectural reasons, structural glued laminated timber (glulam) is widely used. Glulam members are fabricated by multiple layers of relatively thin laminations glued together with moisture-resistant adhesives, creating a large, strong, structural member. These laminations usually are finger-jointed and normally loaded perpendicular to the wide faces of the individual laminations. Higher degree of structural efficiency can be obtained by virtue of wood engineering. Usual thickness of a lamination is 1 or 2 inches for easy seasoning and machine stress rating. Glulam can also be produced in curved shapes, offering extensive design flexibility. In fact, sizes of glulam members are often limited by handling systems and length restrictions imposed by highway transportation systems rather than by the size of the tree.

Layup of Lamination
The use of laminating material can be optimized by selecting and placing the highest quality laminations at critically stressed zones of a cross-section. In addition, the strength-reducing defects may be dispersed throughout the member to achieve higher structural quality of glulam members. It is critically important that glulam members

must be installed with the proper orientation because the laminations are positioned based on the type and intensity of the stresses that it will be subjected to in service.

Balanced layup

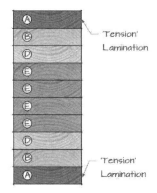

For a beam with balanced layup, the grade of laminations used in a member is symmetrical throughout the depth of the member. The same quality laminations are laid up in a mirror image for the tension and compression zones of a beam. In this way, equal bending capacity in both positive and negative bending can be achieved. This type of member is typically used for continuous multiple span beams or cantilever which can have either the top or bottom of the member stressed in tension.

Unbalanced layup

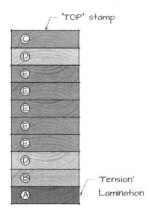

These types of beams are asymmetrical in their layups with the highest quality laminations, referred to as 'tension laminations', used only on the tension zone (bottom) of the member. These are mainly intended for use in simple beams where only the bottom of the beam is subjected to the maximum tensile stress. It is widely accepted that the quality of the laminations used in the tension extreme fiber controls the overall bending strength of the member. Unbalanced beams have the word 'TOP' stamped on the top of the member to ensure the proper installation.

Camber

In addition to aesthetic reasons, there is a structural reason for limiting deflection in large flat roofs. This is to prevent a phenomenon called 'ponding' which may occur when water collects in a depression in a flat roof. As water builds up in a depression, the supporting members will deflect more which causes more water to accumulate.

Once initiated, this vicious cycle can continue until a roof failure occurs. When ponding is a potential problem, beam camber with other positive means to insure proper drainage must be used. Camber is an initial reverse deflection built into glulam members to compensate sagging appearance (deflection) and to avoid ponding effects in flat roof. The glulam industry recommends the use of 1.5 times the calculated dead load deflection for most roof beam applications.

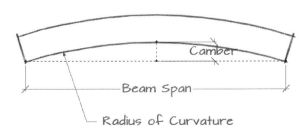

Typical Grade Stamps for Glulam

APA EWS – Quality Control Agency

B – Structural Use

T	:	tension member
C	:	compression member
B	:	simple span bending member
CB	:	continuous or cantilever bending member

IND – Appearance Grade

FRAMING	:	Framing
IND	:	Industrial
IND	:	Industrial
ARCH	:	Architectural
PRE	:	Premium

117-93 – Laminating Specification

24F-1.8E - Combination Symbol
24F : allowable bending stress of 2400 psi
1.8E : MOE of 1.8×10^6 psi)

MILL 0000 - Mill Number

ANSI A190.1-1992 - Standard for Structural Glulam Timber
(ANSI: American National Standards Institute)

9.5 Reference Design Values - TABLE B-4

Bending Combination and Bending about X-X Axis (Major Axis)

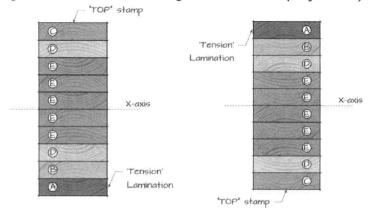

**Bending about X-Axis
Tension Zone in Tension**

**Bending about X-Axis
Compression Zone in Tension**

Bending Combination and Bending about Y-Y Axis (Minor Axis)

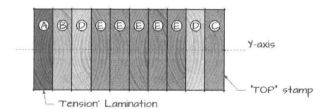

9.6 Adjustment Factors for Glulam

The Glued Laminated Timber (Glulam) Industry provides Reference Design Values that Engineers may use in the design of Glulam Timbers. These Reference Design Values are for `standard conditions'. The standard conditions include physical characteristics of the beam itself and its supports, the nature of the load it is supporting, and other conditions such as temperature, moisture, and so on. Proper design of timber structures requires adjustment of the Reference Design Values to the in-place (in-service or end-use) conditions of the structural member. As discussed in Chapter 1, adjustment factors are used in general for structural wood products such as for dimension lumber, timbers, and structural composites. Here, only noticeable adjustment factors particular to Glulam are discussed below.

Wet Service Factor, C_M

In general wood strength and stiffness properties decrease with increased moisture content. The West Service factor applies generally to other structural wood products as well as for Glulam, though the specific `numbers' and moisture content thresholds differ. For moisture content values of less than 16% the Wet Service factor does not apply (or can be considered Unity). For Glulam timbers with moisture content values 16% or greater the Wet Service factor, $C_M <$ 1.0 as follows:

Wet Service Factors (C_M) for Glulam						Table A-7
use with	F_b	F_t	F_c	F_v	$F_{c\perp}$	E & E_{min}
C_M	0.8	0.8	0.73	0.875	0.53	0.833

Volume Factor, C_V

In sawn lumber, the size effect is taken into account by the size factor C_F. However, full-scale test data indicate that the size effect in glulam is related to the volume of the member rather than to only its depth. Therefore, the volume factor C_V replaces the size factor C_F for use in glulam design. The Bending Design Values for Glulam are based on a standard size of 5.125 in. wide by 12 in. deep by 21 ft long. Beams of greater volume have lesser unit strength in flexural tension, the effect of which is taken into account with the Volume factor, C_V. Note that CV is applied only to glulam beams subjected to loads applied perpendicular to the wide face of the lamination.

$$C_V = \left(\frac{5.125\ in}{b}\right)^{\frac{1}{x}} \left(\frac{12\ in}{d}\right)^{\frac{1}{x}} \left(\frac{21\ ft}{L}\right)^{\frac{1}{x}} \leq 1.0$$

Where,
   b = beam width
   d = beam depth
   L = the length of the beam between points of zero moment
   x = 20 for Glulam manufactured from Southern Pine
      = 10 for Glulam manufactured from all other species

In no case may C_V may be taken to be greater than 1.0. Alternately stated, for all beams up to the `standard' size beam the Volume factor is not applicable. Since the Beam Stability factor addresses the compression zone of the beam, and the Volume factor the tension zone, only the lesser (`controlling') factor of (C_V , C_L) is used to adjust the Bending Design Value in any one application. Only the lesser of applies.

Flat Use Factor, C_{fu}

For Glued Laminated Timbers `laid flat' and for those on edge but loaded from the side the Flat Use factor, C_{fu} is applied to the Bending Design Value F_{by}. It applies only to members less than 12 in. wide (dimension parallel to the wide faces of the laminations) and increases the bending strength. C_{fu} may be obtained in tabular form from the NDS, or may be calculated using the equation below:

$$C_{fu} = \left(\frac{12\ in}{d}\right)^{\frac{1}{9}}$$

where, d is the beam dimension measured parallel to the wide faces of the laminations.

Flat Use Factors (C_{fu}) for Glulam   Table A-13

Member Dimension parallel to wide faces of laminations	C_{fu}
10-3/4" or 10-1/2"	1.01
8-3/4" or 8-1/2"	1.04
6-3/4"	1.07
5-1/8" or 5"	1.10
3-1/8" or 3"	1.16
2-1/2"	1.19

Curvature Factor, C_C

Glued Laminated Timbers may be manufactured in curved geometries. The curved geometry introduces pre-stresses to each lamination. Design Values for Glued Laminated Timbers already accommodate the amounts of pre-stress developed by manufacturing the beams to standard cambers. For more sharply curved beams the Bending Design Value is reduced by the Curvature factor, C_C, as follows:

$$C_C = 1 - 2000 \times \left(\frac{t}{R}\right)^2$$

Where,

t = lamination thickness
R = radius of curvature

For sharply curved beams laminations of smaller thickness are sometimes used to reduce the effect of the pre-stress (increase C_c).

All of the above Adjustment factors apply to the Bending Design Value for Glulam. Some of the factors apply only to Bending (C_L, C_V, and C_{fu}), while others apply to Bending and other design properties (C_D, C_M and C_t), though the actual 'numbers' applied may differ, property to property and even member to members. The following Adjustment factors apply to some of the other Design Values for Glulam.

Shear Reduction Factor, C_{vr}

Shear Parallel to Grain Design Values for Glulam are based on prismatic members subject to static loading. For non-prismatic members (tapered, notched, etc.) and for members subject to cyclic or impact loads the Shear Reduction factor, C_{vr}, is applied to the Shear Parallel to Grain Design Value. Where applicable, C_{vr} = 0.72. This factor appears 'un-named' and in the footnotes of the 2005 NDS and explicitly, as described above, in the 2012 NDS.

Applicability of Adjustment Factors for Glulam Timber   Table A-14

Adjusted Design Values	Base Values		ASD only	ASD and LRFD									LRFD only		
			Load duration factor	Beam stability factor	Volume factor	Temperature factor	Wet service factor	Flat use factor	Curvature factor	Column stability factor	Buckling stiffness factor	Bearing area factor	Format conversion	Resistance factor	Time effect factor
$F_b' = F_b \times$			C_D	C_L	C_V	C_t	C_M	C_{fu}	C_c	–	–	–	K_F	ϕ_b	λ
$F_t' = F_t \times$			C_D	–	–	C_t	C_M	–	C_c	–	–	–	K_F	ϕ_t	λ
$F_v' = F_v \times$			C_D	–	–	C_t	C_M	–	–	–	–	–	K_F	ϕ_v	λ
$F_{c\perp}' = F_{c\perp} \times$			–	–	–	C_t	C_M	–	–	–	–	C_b	K_F	ϕ_c	λ
$F_c' = F_c \times$			C_D	–	–	C_t	C_M	–	–	C_P	–	–	K_F	ϕ_c	λ
$E' = E \times$			–	–	–	C_t	C_M	–	–	–	–	–	–	–	–
$E_{min}' = E_{min} \times$			–	–	–	C_t	C_M	–	–	–	C_T	–	K_F	ϕ_s	–

9.7 Other Engineered wood Products

Laminated Veneer Lumber (LVL)

To the untrained eye, LVL looks like plywood. Thin veneers are glued together like plywood, but the grain of every veneer runs in the same direction. This alignment takes advantage of the natural strength properties of wood. The resulting parallel-laminated lumber outperforms conventional lumber. Production of LVL is more or less a continuous process, so using LVL flanges means that I-Joists can be made in very long lengths. Veneers used to fabricate the LVL are carefully selected. Defects are culled. Strength and stiffness are maximized. Structural values of the LVL flanges are very high.

Oriented Strand Board (OSB)

The finished product has similar properties to plywood, but is uniform and cheaper. It has replaced plywood in many environments, especially the North American structural panel market. The most common uses are as sheathing in walls, floors, and roofs. While OSB does not have a continuous grain like a natural wood, it does have a specific axis of strength. This can be seen by observing the alignment of the surface wood chips. The most accurate method for determining the axis of strength is to examine the ink stamps placed on the wood

by the manufacturer. It is manufactured in wide mats from cross-oriented layers of thin, rectangular wooden strips compressed and bonded together with wax and resin adhesives. The mat is made in a forming line and the layers are built up with the external layers aligned in the panel's strength axis with internal layers cross-oriented. The mat is placed in a thermal press to compress the flakes and bond them by heat activation and curing of the resin that has been coated on the flakes.

Plywood

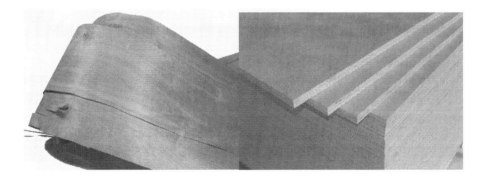

Plywood, a wood structural panel, is sometimes called the original engineered wood product. Plywood is manufactured from sheets of cross-laminated veneer which are peeled from a spinning log and bonded under heat and pressure with durable, moisture-resistant adhesives. By alternating the grain direction of the veneers from layer to layer, or "cross-orienting", panel strength and stiffness in both directions are maximized. Also, cross-graining has several important benefits: it reduces the tendency of wood to split when nailed at the edges; and it reduces expansion and shrinkage, providing improved dimensional stability. There are usually an odd number of plies, so that the sheet is balanced—this reduces warping. A typical plywood panel has face veneers of a higher grade than the core veneers. The principal function of the core layers is to increase the separation between the outer layers where the bending stresses are highest, thus increasing the panel's resistance to bending (moment of inertia). As a result, thicker panels can span greater distances under the same loads. In bending, the maximum stress occurs in the outermost layers, one in tension and the other in compression. Bending stress decreases from the maximum at the face layers to nearly zero at the central layer. Since it is made from whole layers of logs rather than small strands, plywood has a more consistent and less rough appearance than OSB which is another comparable structural wood panel.

Oriented Strand Board vs. Plywood

Oriented strand board (OSB) and plywood are wood structural panels made by compressing and gluing pieces of wood together. While OSB and plywood appear similar and are generally interchangeable, the different ways that each material is manufactured contribute to each having its own unique strengths and weaknesses. OSB is equivalent

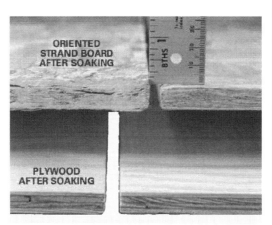

to plywood in strength, is slightly cheaper, but compared to plywood, swells more when it comes into contact with water, especially at panel edges. Building codes and Engineered Wood Association acknowledge that OSB has the same strength and durability as plywood. Some OSB panels have a textured surface, which makes them less slippery when used for roof sheathing. OSB is more uniform, so there are fewer soft spots, such as those that can occur in plywood. But OSB is less stiff than plywood floors by a factor of approximately 10%. In addition, nails and screws are more likely to remain in place more firmly in plywood than in OSB. OSB weighs approximately 15% more than plywood. The heavier weight of OSB makes it harder to install and also put more loads on the house.

Parallel Strand Lumber (PSL)

Parallel strand lumber (PSL) consists of long veneer strands laid in parallel formation and bonded together with an adhesive to form the finished structural section. A strong, consistent material, it has a high load carrying ability and is resistant to seasoning stresses so it is well suited for use as beams and columns for post and beam construction, and for beams, headers, and lintels for light framing construction. PSL is a member of the structural composite lumber (SCL) family of engineered wood products.

I-Joists

An engineered wood joist, more commonly known as an I-joist, is a product designed to improve structural properties of conventional wood joists. The I-joist has very high strength and stiffness for its size and weight. As results, the I-joist is designed to carry heavy loads over long distances while using less lumber than a dimensional solid wood joist of a size necessary to do the same task. I-joists' dimensional stability and little shrinkage help eliminate squeaky floors.

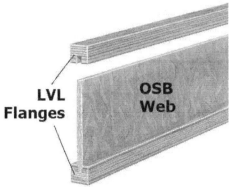

LVL Flanges

OSB Web

An I-joist comprises two main parts; the web and flange. The web is sandwiched between the top and bottom flanges, creating the I-shape. The "I" configuration provides high bending strength and stiffness characteristics due to higher moment of inertia for the weight. The flanges resist bending stresses, and the web provides shear performance. The flange is usually made from laminated veneer lumber (LVL) finger-jointed together for ultimate strength. Production of LVL is more or less a continuous process, so using LVL flanges means that I-Joists can be made in very long lengths. It is then grooved on one side to receive the web. The web is typically made from plywood or oriented strand board (OSB). According to The Engineered Wood Association (APA), up to half the homes built in the United States now use engineered wood I-beams. I-joists require correct installation. The biggest negative to using I-joists is the special design and layouts required. Typically, the company supplying the I-joists creates a special, engineered layout showing sizes, spacing and connections. The supplier will typically include all special connectors and beams within the floor. Engineered wood I-joists are extremely prone to failure because of fire damage. Unlike with sawn lumber, the center section is super thin, so it has little mass to resist fire. Additionally, the heat from a fire may melt the glue holding the plies of the center section together, causing disintegration that weakens the joist. Once the web is burnt through, the outer flanges don't have the strength to hold up the floor. This can create dangerous situations for fire professionals when they fail sooner than expected.

Sawn Lumber Beam vs. I-Joists

More than 70% of all floors are framed with dimension lumber. Performance of I-Joists is clearly superior to that provided by dimension lumber. Consistency of the material quality is the key advantage of engineered wood. Fabricated from dry materials to very tight manufacturing standards, wood I-joists generally do not shrink, warp, cup, crown, or twist.

I-Joists provide:

1. better dimensional stability - little shrinkage
2. improved strength and stiffness
3. lighter weight to handle
4. webs that are easier to drill for HVAC, plumbing, and electrical
5. design flexibility with increased span potential
6. increased on-center spacing.

High prices and unfamiliarity with a new product have kept I-joists away from most job sites. Until recently, it was difficult for I-joists to compete with sawn lumber on starter homes and houses with a basic design. I-Joists are becoming more familiar and builders less intimidated. Dimension lumber joists are typically available in lengths up to 16-feet and installed at 16-inch on-center spacing. If a builder uses I-joists the number of joists is instantly cut in half. The reduction in the number of joists translates to less nailing of decks and rim joists. I-joists are available in lengths up to 60 feet, so reaching across the house from sill-to-sill is not a problem.

GLOSSARY

Finger-Jointed Lumber
Short lengths of wood that have been machined with finger joints at the ends so they mate perfectly when glued together. Finger jointing adds surface area to the joint which greatly increases its strength. By salvaging short, knot-free lengths from low quality lumber to make long lengths of higher grade lumber, finger jointed wood reduces waste and adds to the sustainability and renewability of the wood.

Glulam
Glue-laminated timber refers to large, structural members made by gluing together pieces of dimension lumber. It's a unique structural material because it can be formed into many curved shapes and the sizes are limited only by transportation restrictions. It is generally used for columns and beams, and frequently for curved members loaded in combined bending and compression. Glulam also is used as an exposed architectural product or it can be hidden or left unfinished to only serve a structural role.

I-Joists
Wood I-joists are a structural engineered wood product often used for joists and rafters in residential and commercial construction. I-joists are made by gluing solid sawn lumber or laminated veneer lumber (LVL) flanges to a web made of plywood or oriented strand board (OSB). I-joists exhibit uniform stiffness, strength and are lightweight.

Laminated Veneer Lumber (LVL)
LVL has several layers of wood veneers and adhesive. First used during World War II to make airplane propellers, it's been used since the mid-1970's to build beams and headers for buildings where high strength, stability and reliability are required.

Machine Stress Rated Lumber
Standard lumber is load rated based on dimension and visual inspection. Engineered lumber is also rated according to its design strength. But the only way to be absolutely certain that a given piece of lumber will meet its rating is to stress it mechanically. Machine Stress Rated (MSR) lumber is just that: lumber that has been mechanically stressed to its rated load before it is used in a project. The benefit is lumber that is less likely to fail mechanically once installed. This results in fewer expensive onsite repairs.

Medium Density Fiberboard (MDF)
A wood-based panel made from fine cellulose fibers combined with a synthetic resin or other suitable bonding system, which is then joined together under heat and pressure. Because of the fineness of the fibers, it forms and machines smoothly and precisely, making it an ideal substrate for thin laminates.

Oriented Strand Board (OSB)
Wood panel used as a sheathing material for floors, walls and roofs. OSB has grown enormously in popularity over the past decade because it's economical and has excellent racking resistance, which is, the ability to strengthen a building against horizontal loads such

as extreme winds or the forces of earthquakes.

Parallel Strand Lumber

PSL is comparable in strength to laminated veneer lumber. Both products are often grouped under the heading Structural Composite Lumber (SCL). PSL is made from long strands of wood chopped from wood veneer. This process allows the removal of growth defects from the raw material. PSL is then made into large billets and sawn into stock and custom sizes.

Particleboard

A panel product made from sawdust and other residue left over from the manufacture of lumber and other wood products. Since it's derived from other wood products, particleboard is an environmentally friendly choice.

Plywood

An engineered panel product consisting of veneers glued together so the grain direction of the ply is perpendicular to that of adjacent plies. This gives plywood equal strength in its length and width.

Rim Board

A board used to surround I-joist floor-joists. Rim boards mount between sills and wall plates. In addition to transferring lateral and vertical loads to the frame of the house, RIM Board can be ordered to match the depth of I-Joist framing members—something that conventional lumber typically does not match.

Structural plywood

Used for construction and industrial applications, made with waterproof glue to maintain strength in exterior applications.

9.8 Workshop for **Glulam Beam Analysis**

Case Study 9-1a

A *Western Species* glulam beam spans 24 ft. and is spaced at 14 ft o.c. The beam supports a dead load of 15 psf (including self-weight of the glulam beam) and a snow load of 35 psf. The gluelam beam size is 5 ½ × 19 ½ and stress class is 20F-1.5E. The bending is about its strong axis and the tension zone is in tension. The live load deflection limit is L/240 and the total load deflection is L/180. Determine the adequacy of the glulam beam. Also, determine the proper size of camber of the beam.

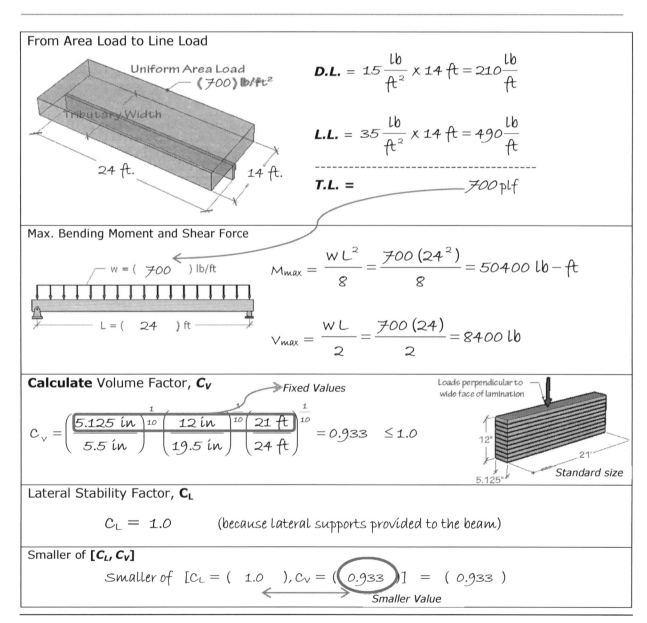

From Area Load to Line Load

Uniform Area Load
(700) lb/ft²

Tributary Width

24 ft.   14 ft.

$$D.L. = 15\frac{lb}{ft^2} \times 14\,ft = 210\frac{lb}{ft}$$

$$L.L. = 35\frac{lb}{ft^2} \times 14\,ft = 490\frac{lb}{ft}$$

- -

T.L. = _____ 700 plf

Max. Bending Moment and Shear Force

w = (700) lb/ft

L = (24) ft

$$M_{max} = \frac{wL^2}{8} = \frac{700\,(24^2)}{8} = 50400\ lb-ft$$

$$V_{max} = \frac{wL}{2} = \frac{700\,(24)}{2} = 8400\ lb$$

Calculate Volume Factor, C_V → Fixed Values

Loads perpendicular to wide face of lamination

$$C_v = \left(\frac{5.125\ in}{5.5\ in}\right)^{\frac{1}{10}} \left(\frac{12\ in}{19.5\ in}\right)^{\frac{1}{10}} \left(\frac{21\ ft}{24\ ft}\right)^{\frac{1}{10}} = 0.933 \leq 1.0$$

12"
5.125"
21'
Standard size

Lateral Stability Factor, C_L

$C_L = 1.0$ (because lateral supports provided to the beam)

Smaller of [C_L, C_V]

Smaller of [C_L = (1.0), C_V = (0.933)] = (0.933)

← _____ → Smaller Value

Adjusted Design Values

	Ref. Value	C_D	C_M	C_t	Smaller [C_L, C_V]	Allowable Value
$F_b' =$	2000 psi	1.15	1.0	1.0	0.933	2147 psi
$F_v' =$	210 psi	1.15	1.0	1.0	N/A	242 psi
$E' =$	1500 ksi	N/A	1.0	1.0	N/A	1500 ksi

Section Properties of Beam

$$S = \frac{b(d)^2}{6} = \frac{5.5(19.5)^2}{6} = 349 \text{ in}^3 \qquad I = \frac{b(d)^3}{12} = \frac{5.5(19.5)^3}{12} = 3398 \text{ in}^4$$

Maximum Bending Stress

$$(f_b)_{max} = \frac{M_{max}}{S} = \frac{50400 \text{ lb}-\text{ft}(12)}{349 \text{ in}^3} = 1732 \frac{\text{lb}}{\text{in}^2}$$

check if $f_b \leq F_b'$ 1732 psi < 2147 psi   (O.K.) or N.G.)

Maximum Shear Stress

$$(f_v)_{max} = 1.5 \times \frac{V}{A} = 1.5 \times \frac{8400}{5.5(9.5)} = 117 \text{ psi}$$

check if $f_v \leq F_v'$ 117 psi < 242 psi   (O.K.) or N.G.)

Deflection Check

$$\Delta_{total} = \frac{5\, W_{total}\, L^4 (12^3)}{384\, E'\, I} = \frac{5(700)(24)^4(12^3)}{384(1.5 \times 10^6)(3398)} = 1.03 \text{ in} \leq \frac{L}{180} = \frac{24(12)}{180} = 1.6 \text{ in}$$

(O.K.) or N.G.)

$$\Delta_{live} = \Delta_{total} \times \frac{W_{live}}{W_{total}} = 1.03 \text{ in} \times \frac{35}{45} = 0.80 \text{ in} \leq \frac{L}{240} = \frac{24(12)}{240} = 1.2 \text{ in}$$

(O.K.) or N.G.)

$$\Delta_{dead} = \Delta_{total} \times \frac{W_{dead}}{W_{total}} = 1.03 \times \frac{10}{45} = 0.23 \text{ in}$$

$$\text{Camber} = (1.5)\, \Delta_D = 1.5\,(0.23 \text{ in}) = 0.345 \text{ in.}$$

Case Study 9-1b

A *Western Species* glulam beam spans 28 ft. and is spaced at 16 ft o.c. The beam supports a dead load of 20 psf (including self-weight of the glulam beam) and a snow load of 40 psf. The gluelam beam size is 5 × 24 ¾ and stress class is 24F-1.8E. The bending is about its strong axis and the tension zone is in tension. Determine the adequacy of the glulam beam. The live load deflection limit is L/240 and the total load deflection is L/180.

From Area Load to Line Load
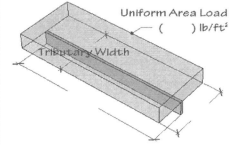 Uniform Area Load ( ) lb/ft² Tributary Width

Max. Bending Moment and Shear Force
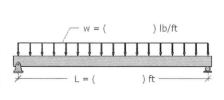 w = ( ) lb/ft L = ( ) ft $$M_{max} = \frac{wL^2}{8} =$$ $$V_{max} = \frac{wL}{2} =$$

Calculate Volume Factor, C_V
$$C_v = \left(\frac{5.125 \ in}{}\right)^{\frac{1}{10}} \left(\frac{12 \ in}{}\right)^{\frac{1}{10}} \left(\frac{21 \ ft}{}\right)^{\frac{1}{10}} = (\qquad) \le 1.0$$ $$C_v =$$ 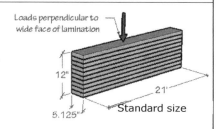 Loads perpendicular to wide face of lamination 12" 21' 5.125" Standard size

Lateral Stability Factor, C_L
$$C_L =$$

Smaller [C_L, C_V]
Smaller [$C_L = (\qquad)$, $C_V = (\qquad)$] = ($\qquad$)

Adjusted Design Values

	Ref. Value	C_D	C_M	C_t	Smaller [C_L, C_V]	Allowable Value
$F_b' =$						
$F_v' =$					N/A	
$E' =$		N/A			N/A	

Section Properties of Beam

$$S = \qquad\qquad I = \qquad\qquad E =$$

Maximum Bending Stress

$$(f_b)_{max} = \frac{M_{max}}{S} =$$

$$\text{check if} \quad f_b \leq F_b' \qquad\qquad\qquad (O.K. \ or \ N.G.)$$

Maximum Shear Stress

$$(f_v)_{max} = 1.5 \frac{V_{max}}{A} =$$

$$\text{check if} \quad f_v \leq F_v' \qquad\qquad\qquad (O.K. \ or \ N.G.)$$

Deflection Check

$$\Delta_{total} = \frac{5 \, W_{total} \, L^4 (12^3)}{384 \, E' \, I} = \qquad\qquad\qquad \leq \frac{L}{180} =$$

$$(O.K. \ or \ N.G.)$$

$$\Delta_{live} = \Delta_{total} \times \frac{W_{live}}{W_{total}} = \qquad\qquad\qquad \leq \frac{L}{240} =$$

$$(O.K. \ or \ N.G.)$$

$$\Delta_{dead} = \Delta_{total} \times \frac{W_{dead}}{W_{total}} =$$

$$Camber = (1.5) \, \Delta_D =$$

Workshop 9-1a

A *Western Species* glulam beam spans 40 ft and is spaced 14 ft. on center. The beam supports a dead load of 15 psf (including self-weight of the glulam beam) and a roof live load of 30 psf. The gluelam beam size is 8 ½ × 22 and stress class is 24F-1.7E. The bending is about its strong axis and the tension zone is in tension. Determine the adequacy of the

glulam beam. The compression side of the beam is laterally supported by the roof systems. The live load deflection limit is L/240 and the total load deflection is L/180.

From Area Load to Line Load
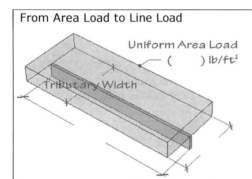

Max. Bending Moment and Shear Force
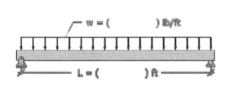 $M_{max} = \dfrac{w\,L^2}{8} =$ $V_{max} = \dfrac{w\,L}{2} =$

Volume Factor, C_V
$C_v = \left(\dfrac{5.125\ in}{}\right)^{\frac{1}{10}} \left(\dfrac{12\ in}{}\right)^{\frac{1}{10}} \left(\dfrac{21\ ft}{}\right)^{\frac{1}{10}} = (\qquad) \le 1.0$ $C_V =$

Lateral Stability Factor, C_L
$C_L =$

Smaller of $[C_L, C_V]$
Smaller of $[C_L = (\qquad), C_V = (\qquad)] = (\qquad)$

Adjusted Design Values

	Ref. Value	C_D	C_M	C_t	Smaller [C_L, C_V]	Allowable Value
$F_b' =$						
$F_v' =$					N/A	
$E' =$		N/A			N/A	

Section Properties of Beam

$$S = \qquad\qquad I = \qquad\qquad E =$$

Maximum Bending Stress

$$(f_b)_{max} = \frac{M_{max}}{S} =$$

$$\text{Check if} \quad f_b \leq F_b' \qquad\qquad\qquad\qquad (O.K. \; or \; N.G.)$$

Maximum Shear Stress

$$(f_v)_{max} = 1.5\, \frac{V_{max}}{A} =$$

$$\text{Check if} \quad f_v \leq F_v' \qquad\qquad\qquad\qquad (O.K. \; or \; N.G.)$$

Deflection Check

$$\Delta_{total} = \frac{5\, W_{total}\, L^4\, (12^3)}{384\, E'\, I} = \qquad\qquad\qquad \leq \quad \frac{L}{180} =$$

$$(O.K. \; or \; N.G.)$$

$$\Delta_{live} = \Delta_{total} \times \frac{W_{live}}{W_{total}} = \qquad\qquad\qquad \frac{L}{240} = \quad \leq$$

$$(O.K. \; or \; N.G.)$$

$$\Delta_{dead} = \Delta_{total} \times \frac{W_{dead}}{W_{total}} =$$

$$\text{Camber} = (1.5)\, \Delta_D =$$

Workshop 9-1b

A *Western Species* gluelam beam spans 35 ft and is spaced 10 ft. on center. The beam supports a dead load of 20 psf (including self-weight of the glulam beam) and a snow of 40 psf. The gluelam beam size is 5 ½× 23 ⅜ and stress class is 24F-1.8E. The bending is about its strong axis and the tension zone is in tension. Determine the adequacy of the glulam beam. The live load deflection limit is L/240 and the total load deflection is L/180.

Workshop 9-1c

A gluelam beam spans 28 ft and is spaced 20 ft. on center. The beam supports a dead load of 20 psf (including self-weight of the glulam beam) and a snow of 35 psf. The gluelam beam size is 6 ¾ × 24 ¾ and stress class is 24F-1.7E. The bending is about its strong axis and the tension zone is in tension. Determine the adequacy of the glulam beam.

9.9 Workshop for **Glulam Beam Design**

Case Study 9-2

Select the Western Species gluelam sizes for the typical members, *Girders 1* and *Beams 1* of Ballard Library in Seattle. The roof system supports a dead load of 20 lb/ft^2 and a snow load of 40 lb/ft^2. The stress class of the gluelam members is **24F-1.8E**. The bending is about its strong axis and the tension zone is in tension. As shown in the picture, the roof diaphragm provides a full lateral support to the compression side of the girders and beams. The cantilever portion of the over-hang beam is not loaded so that it can be neglected in the analysis.

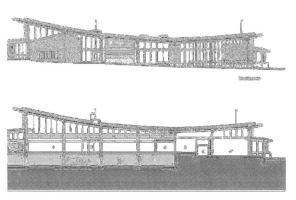

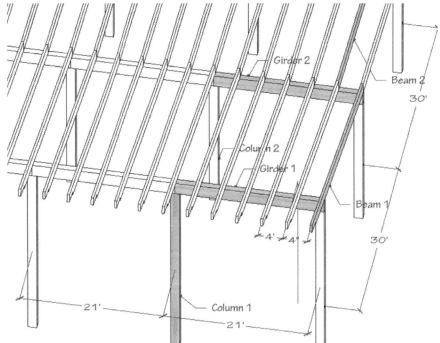

Workshop 8-2

Select the Western Species gluelam sizes for the typical members, *Girders 2* and *Beams 2* of Ballard Library in Seattle. The roof system supports a dead load of 18 lb/ft^2 and a snow load of 35 lb/ft^2. The stress class of the gluelam members is **20F-1.5E**. The roof diaphragm provides a full lateral support to the compression side of the girders and beams.

Case Study 9-2a   Design of (*Girder 1*)

From Area Load to Line Load

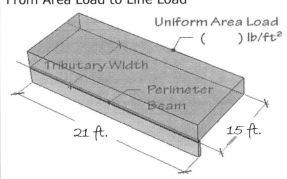

Uniform Area Load
(   ) lb/ft²

Tributary Width

Perimeter Beam

21 ft.   15 ft.

$$D.L. = 20\frac{lb}{ft^2} \times 15 \, ft = 300 \frac{lb}{ft}$$

$$L.L. = 40\frac{lb}{ft^2} \times 15 \, ft = 600 \frac{lb}{ft}$$

$$T.L. = \qquad\qquad 900 \ plf$$

Max. Bending Moment and Shear Force

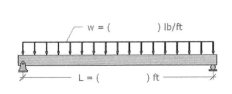

w = (   ) lb/ft

L = (   ) ft

$$M_{max} = \frac{wL^2}{8} = \frac{900\,(21^2)}{8} = 46123 \, lb-ft$$

$$V_{max} = \frac{wL}{2} = \frac{900\,(21)}{2} = 9450 \, lb$$

Assume Volume Factor, C_V

    Assume $C_V = 0.92$ (*size of the beam is unknown at this stage of design*)

Lateral Stability Factor, C_L

    $C_L = 1.0$ (*because lateral supports are provided to the beam*)

Smaller of [C_L, C_V]

    Smaller of $[C_L = (\ 1.0 \ \ ), C_V = (\ 0.93 \ )]$ =   (0.93)

Adjusted Design Values

	Ref. Value	C_D	C_M	C_t	Smaller [C_L, C_V]	Allowable Value
$F_b' =$	2400 psi	1.15	1.0	1.0	0.92	2539 psi
$F_v' =$	265 psi	1.15	1.0	1.0	N/A	305 psi
$E' =$	1800 ksi	N/A	1.0	1.0	N/A	1800 ksi

Required Section Modulus

$$S_{requird} = \frac{M_{max}}{F'_b} = \frac{46123 \, lb-ft\,(12)}{2539 \, \frac{lb}{in^2}} = 234.5 \, in^3$$

Selection of Trial Size (from APA Table 5) *with required section modulus of 234.5 in³*

Beam Designation	S (in³)	A (in²)	Moment Capacity (lb-ft)	Shear Capacity (lb)	EI (10⁶ lb-in²)	Selection
3 ½ X 21	257	73.5	51450	12985	4862	Try this
5 1/8 X 18	277	92.3	55350	16298	4483	
5 ½ X 16	235	88.0	46933	15547	3379	

Calculate Volume Factor, C_V

$$C_V = \left(\frac{5.125 \text{ in}}{3.5 \text{ in}}\right)^{\frac{1}{10}} \left(\frac{12 \text{ in}}{21 \text{ in}}\right)^{\frac{1}{10}} \left(\frac{21 \text{ ft}}{21 \text{ ft}}\right)^{\frac{1}{10}} = (0.982) \leq 1.0$$

Adjust Member Capacity with Calculated C_V-value

	Ref. Value	C_D	C_M	C_t	Smaller $[C_L, C_V]$	Adjusted Value
$M' =$	51450 lb-in	1.15	1.0	1.0	0.982	58102 lb-in
$V' =$	12985 lb	1.15	1.0	1.0	N/A	14933 lb

$M_{max} = 46123 \text{ lb-ft} < M_{allowable} = 58102 \text{ lb-ft}$ (O.K.) or N.G.)

$V_{max} = 9450 \text{ lb} < V_{allowable} = 14933 \text{ lb}$ (O.K.) or N.G.)

Deflection Check

$$\Delta_{total} = \frac{5 \, w_{total} \, L^4 (12^3)}{384 \, E' \, I} = \frac{5 \, (900)(21)^4 (12^3)}{384 \, (4862 \times 10^6)} = 0.81 \text{ in} \quad \leq \quad \frac{L}{180} = \frac{21(12)}{180} = 1.4 \text{ in}$$

(O.K.) or N.G.)

$$\Delta_{live} = \Delta_{total} \times \frac{w_{live}}{w_{total}} = 0.81 \text{ in} \times \frac{600 \text{ lb/ft}}{900 \text{ lb/ft}} = 0.54 \text{ in} \quad \leq \quad \frac{L}{240} = \frac{21(12)}{240} = 1.05 \text{ in}$$

(O.K.) or N.G.)

$$\Delta_{dead} = \Delta_{total} \times \frac{w_{dead}}{w_{total}} = 0.81 \text{ in} \times \frac{300 \text{ lb/ft}}{900 \text{ lb/ft}} = 0.27 \text{ in}$$

Suggested Camber = $(1.5) \, \Delta_D = 1.5 \, (0.27 \text{ in}) = 0.41 \text{ in.}$

Case Study 9-2b Design of (Beam 1)

From Area Load to Line Load

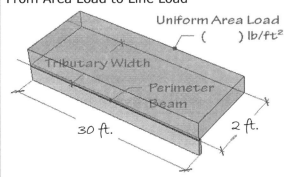

$D.L. = 20\dfrac{lb}{ft^2} \times 2\,ft = 40\dfrac{lb}{ft}$

$L.L. = 40\dfrac{lb}{ft^2} \times 2\,ft = 80\dfrac{lb}{ft}$

$T.L. =$ 120 plf

Max. Bending Moment and Shear Force

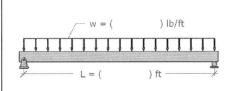

w = ( ) lb/ft

L = ( ) ft

$M_{max} = \dfrac{wL^2}{8} = \dfrac{120\,(30^2)}{8} = 13500\ lb-ft$

$V_{max} = \dfrac{wL}{2} = \dfrac{120\,(30)}{2} = 1800\ lb$

Assume Volume Factor, C_V

Assume $C_V = 0.95$ (size of the beam is unknown at this stage of design)

Lateral Stability Factor, C_L

$C_L = 1.0$ (because lateral supports are provided to the beam)

Smaller of $[C_L, C_V]$

Smaller of $[C_L = (\ 1.0\ ), C_V = (\ 0.95\ )] = (\ 0.95\ )$

Adjusted Design Values

	Ref. Value	C_D	C_M	C_t	Smaller $[C_L, C_V]$	Allowable Value
$F_b' =$	2400 psi	1.15	1.0	1.0	0.95	2622 psi
$F_v' =$	265 psi	1.15	1.0	1.0	N/A	305 psi
$E' =$	1800 ksi	N/A	1.0	1.0	N/A	1800 ksi

Required Section Modulus

$S_{requird} = \dfrac{M_{max}}{F'_b} = \dfrac{13500\ lb-ft\,(12)}{2622\ \dfrac{lb}{in^2}} = 61.78\ in^3$

Selection of Trial Size (from APA Table 5) *with required section modulus of 55.86 in³*

Beam Designation	S (in³)	A (in²)	Moment Capacity (lb-ft)	Shear Capacity (lb)	EI (10⁶ lb-in²)	Selection
3 1/8 X 12	75	37.5	15000	6625	810	
3 ½ X 10 ½	64.0	36.8	12863	6493	608	Try this

Calculate Volume Factor, C_V

$$C_V = \left(\frac{5.125 \text{ in}}{3.125 \text{ in}}\right)^{\frac{1}{10}} \left(\frac{12 \text{ in}}{18 \text{ in}}\right)^{\frac{1}{10}} \left(\frac{21 \text{ ft}}{30 \text{ ft}}\right)^{\frac{1}{10}} = (1.02) \quad \leq 1.0$$

Adjust Member Capacity with Calculated C_V

	Ref. Value	C_D	C_M	C_t	Smaller [C_L, C_V]	Adjusted Value
M' =	15000 lb-in	1.15	1.0	1.0	1.0	17250 lb-in
V' =	6493 lb	1.15	1.0	1.0	N/A	6667 lb

$$M_{max} = 13500 \text{ lb-ft} \quad < \quad M_{allowable} = 17250 \text{ lb-ft} \qquad (\text{O.K.}) \text{ or N.G.})$$

$$V_{max} = 1800 \text{ lb} \quad < \quad V_{allowable} = 6667 \text{ lb} \qquad (\text{O.K.}) \text{ or N.G.})$$

Deflection Check

$$\Delta_{total} = \frac{5 \, W_{total} \, L^4 (12^3)}{384 \, E \, I} = \frac{5 (120)(30)^4 (12^3)}{384 (608 \times 10^6)} = 3.6 \text{ in} \qquad \not\leqslant \qquad \frac{L}{180} = \frac{30(12)}{180} = 2.0 \text{ in}$$

$$(\text{O.K. or } \boxed{N.G.})$$

2nd Design Cycle

Note: The beam 3 ½ x 10 ½ is found to be strong enough but not stiff enough. A new design cycle must be entered but only deflection check is needed.

Required beam stiffness – (EI value)

$$EI_{required} = EI_{existing} \times \frac{\Delta_{total}}{\Delta_{limit}} = 608 \, (10^6) \text{ in}^2 \times \frac{3.6 \text{ in}}{2.0 \text{ in}} = 1094 \, (10^6) \text{ in}^2$$

Selection of Trial Size based on beam stiffness with required EI of $1094 (10^6)$ in^2

Beam Designation	S (in³)	A (in²)	Moment Capacity (lb-ft)	Shear Capacity (lb)	EI (10^6 lb-in²)	Selection
3 1/8 X 13 1/2		42.2			1153	Try this
3 1/2 X 13 1/2		47.3			1292	

Deflection Check

$$\Delta_{total} = \frac{5\,W_{total}\,L^4\,(12^3)}{384\,E\,I} = \frac{5\,(120)\,(30)^4\,(12^3)}{384\,(1153 \times 10^6)} = 1.9 \text{ in} \qquad \leq \qquad \frac{L}{180} = \frac{30(12)}{180} = 2.0 \text{ in}$$

(O.K. or N.G.)

$$\Delta_{live} = \Delta_{total} \times \frac{W_{live}}{W_{total}} = 1.9 \text{ in} \times \frac{80 \text{ lb/ft}}{120 \text{ lb/ft}} = 1.26 \text{ in} \qquad \frac{L}{240} = \frac{30(12)}{240} = 1.5 \text{ in} \quad \leq$$

(O.K. or N.G.)

$$\Delta_{dead} = \Delta_{total} \times \frac{W_{dead}}{W_{total}} = 1.9 \text{ in} \times \frac{40 \text{ lb/ft}}{120 \text{ lf/ft}} = 0.63 \text{ in}$$

Suggested Camber = $(1.5)\,\Delta_D = 1.5\,(0.63 \text{ in}) = 0.95$ in.

Workshop 8-2a   Design of ($Girder\ 2$)

From Area Load to Line Load

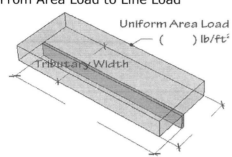

Uniform Area Load
(   ) lb/ft²

Tributary Width

D.L. =

L.L. =

T.L. =

Max. Bending Moment and Shear Force

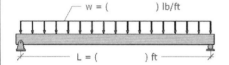

w = (   ) lb/ft

L = (   ) ft

$M_{max} =$

$V_{max} =$

Assume C_V-value

Assume $C_V =$ (size of the beam is unknown at this stage of design)

C_L-value

$C_L =$ (Check if lateral supports provided to the beam)

Smaller [C_L, C_V]

Smaller [C_L = (   ), C_V = (   )] = (   )

Adjusted Design Values

	Ref. Value	C_D	C_M	C_t	Smaller [C_L, C_V]	Allowable Value
$F_b' =$						
$F_v' =$					N/A	
$E' =$		N/A			N/A	

Required Section Modulus

$$S_{requird} = \frac{M_{max}}{F'_b} =$$

Selection of Trial Size (from APA Table 5) *with required section modulus of* ( )

Beam Designation	S (in³)	A (in²)	Moment Capacity (lb-ft)	Shear Capacity (lb)	EI (10⁶ lb-in²)	Selection

Calculate C_V-value

$$C_v = \left(\frac{5.125 \text{ in}}{}\right)^{\frac{1}{10}} \left(\frac{12 \text{ in}}{}\right)^{\frac{1}{10}} \left(\frac{21 \text{ ft}}{}\right)^{\frac{1}{10}} = (\quad) \leq 1.0$$

Adjust Member Capacity with Calculated C_V-value

	Ref. Values	C_D	C_M	C_t	Smaller $[C_L, C_V]$	Adjusted Values
$M' =$						
$V' =$					N/A	

$M_{max} = ($) $<$ $M_{allowable} = ($) (O.K. or N.G.)

$V_{max} = ($) $<$ $V_{allowable} = ($) (O.K. or N.G.)

Deflection Check

$$\Delta_{total} = \frac{5 \, W_{total} \, L^4 \, (12^3)}{384 \, E' \, I} = \qquad\qquad \leq \frac{L}{180} = n$$

(O.K. or N.G.)

$$\Delta_{live} = \Delta_{total} \times \frac{W_{live}}{W_{total}} = \qquad\qquad \leq \frac{L}{240} =$$

(O.K. or N.G.)

$$\Delta_{dead} = \Delta_{total} \times \frac{W_{dead}}{W_{total}} =$$

Suggested Camber $= (1.5) \, \Delta_D =$

Workshop 8-2b   Design of (Beam 2)

From Area Load to Line Load

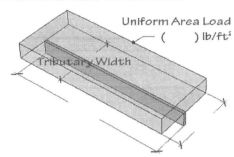

Uniform Area Load
(   ) lb/ft²

Tributary Width

D.L. =

L.L. =

T.L. =

Max. Bending Moment and Shear Force

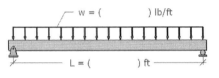

w = (   ) lb/ft

L = (   ) ft

$M_{max} =$

$V_{max} =$

Assume C_V-value

Assume $C_V =$ (size of the beam is unknown at this stage of design)

C_L-value

$C_L =$ (Check if lateral supports provided to the beam)

Smaller [C_L, C_V]

Smaller   [C_L = (   ), C_V = (   )]   = (   )

Adjusted Design Values

	Ref. Value	C_D	C_M	C_t	Smaller [C_L, C_V]	Allowable Value
$F_b' =$						
$F_v' =$					N/A	
$E' =$		N/A			N/A	

Required Section Modulus

$$S_{requird} = \frac{M_{max}}{F'_b} =$$

Selection of Trial Size (from APA Table 5) *with required section modulus of* ( )

Beam Designation	S (in³)	A (in²)	Moment Capacity (lb-ft)	Shear Capacity (lb)	EI (10⁶ lb-in²)	Selection

Calculate C$_V$-value

$$C_V = \left(\frac{5.125 \text{ in}}{} \right)^{\frac{1}{10}} \left(\frac{12 \text{ in}}{} \right)^{\frac{1}{10}} \left(\frac{21 \text{ ft}}{} \right)^{\frac{1}{10}} = (\qquad) \leq 1.0$$

Adjust Member Capacity with Calculated C$_V$-value

	Ref. Value	C_D	C_M	C_t	Smaller [C_L, C_V]	Adjusted Value
M' =						
V' =					N/A	

$$M_{max} = (\qquad) < M_{allowable} = (\qquad) \qquad (O.K. \text{ or } N.G.)$$

$$V_{max} = (\qquad) < V_{allowable} = (\qquad) \qquad (O.K. \text{ or } N.G.)$$

Deflection Check

$$\Delta_{total} = \frac{5 \, W_{total} \, L^4 \, (12^3)}{384 \, E' \, I} = \qquad\qquad \leq \frac{L}{180} = n$$

$$(O.K. \text{ or } N.G.)$$

$$\Delta_{live} = \Delta_{total} \times \frac{W_{live}}{W_{total}} = \qquad\qquad \leq \frac{L}{240} =$$

$$(O.K. \text{ or } N.G.)$$

$$\Delta_{dead} = \Delta_{total} \times \frac{W_{dead}}{W_{total}} =$$

$$\text{Suggested Camber} = (1.5) \, \Delta_D =$$

9.10 Workshop for **Glulam Column Analysis**

Case Study 8-3a

    Determine the axial compression load capacity of the Western Species glulam Column 1 in the figure shown below. The column has a cross-section of 6 ¾ x 11 and is continuous to from the ground floor to the rafter. The stress class of the gluelam members is **24F-1.7E**. MC will exceed 16 percent. Normal temperatures apply. The critical load combination has been found to be D.L + S.L.

Case Study 8-3a

    Determine the axial compression load capacity of the Western Species glulam Column 2 in the figure shown below. The column has a cross-section of 5 ½ x 9 and is continuous to from the ground floor to the rafter. The stress class of the gluelam members is **24F-1.8E**. MC will exceed 16 percent. Normal temperatures apply. The critical load combination has been found to be D.L + S.L.

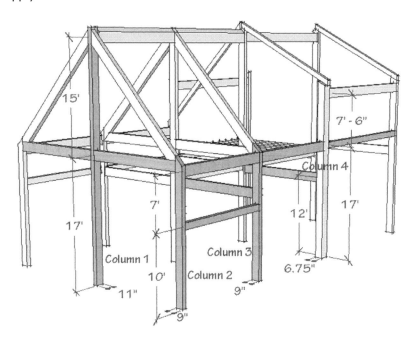

Workshop 8-3a

Determine the axial compression load capacity of the Western Species glulam Column 3 in the figure shown above. The column has a cross-section of 5 ½ x 9 and is continuous to from the ground floor to the rafter. The stress class of the gluelam members is **24F-1.8E**. MC will exceed 16 percent. Normal temperatures apply. The critical load combination has been found to be D.L + S.L.

Workshop 8-3b

    Determine the axial compression load capacity of the Western Species glulam Column 4 in the figure shown above. The column has a cross-section of 6 ¾ x 9 and is continuous to from the ground floor to the rafter. The stress class of the gluelam members is **24F-1.8E**. MC will not exceed 16 percent. Normal temperatures apply. The critical load combination has been found to be D.L + L.L.

Case Study 8-3a Determine the axial compression load capacity of the Western Species glulam Column 1 in the figure sawn below. The column has a cross-section of 6 ¾ x 11 and is continuous to from the ground floor to the rafter. The stress class of the gluelam members is **24F-1.7E**. MC will exceed 16 percent. Normal temperatures apply. The critical load combination has been found to be D.L + S.L.

Section Properties

b = 6.75 in	d = 11 in	A = 74.25 in²

Adjusted Design Values

	Ref. Values	C_D	C_M	C_t	Allowable Values
$F_c{}^* =$	1000 psi	1.15	0.73	1.0	839.5 psi
$E'_{x\,min} =$	880000 psi	N/A	0.833	1.0	733040 psi
$E'_{y\,min} =$	670000 psi	N/A	0.833	1.0	558110 psi

Note that the values of Modulus of Elasticity are not the same in x- and y-direction.

X-axis Buckling

x-plane

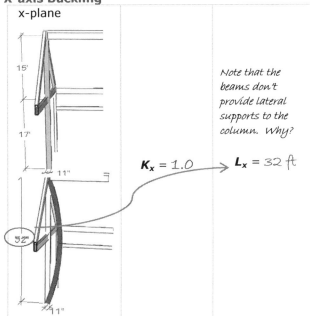

Note that the beams don't provide lateral supports to the column. Why?

$K_x = 1.0$ → $L_x = 32$ ft

$$\left(\frac{Kl}{d}\right)_x = \frac{(1.0)\,(32 \times 12)\,in}{11\,in} = 34.91$$

Design Parameters

$$F_c = \frac{0.822\,E'_{x\,min}}{\left(\frac{Kl}{d}\right)_x^2} = \frac{0.822\,(733040\,psi)}{(34.91)^2} = 494.42\,psi$$

$$\frac{F_{cE}}{F_c{}^*} = \frac{494.42\,psi}{839.5\,psi} = 0.589$$

Column Stability Factor, C_p $C_p = 0.5294$ (Use *Appendix, Table B-6 and Linear Interpolation*)

Allowable Compressive Stress $F_c' = F_c{}^* \times (C_P) = 839.5\,psi \times 0.5294 = 444.43\,psi$

Allowable Compressive Load for x-axis buckling, $P' = F_c'(A) = 444.45\,psi\ (74.25\,in^2) = 32999\,lb$

Y-axis Buckling

y-plane

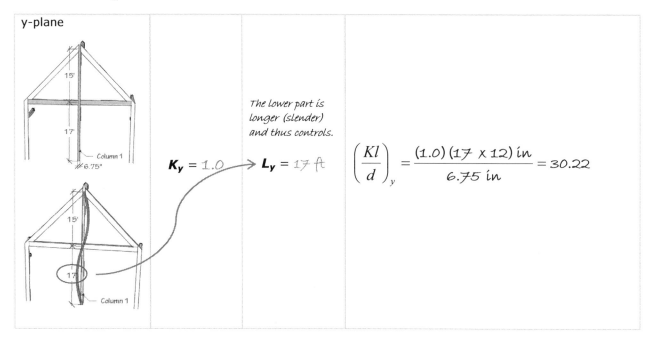

The lower part is longer (slender) and thus controls.

$K_y = 1.0 \rightarrow L_y = 17 \text{ ft}$

$$\left(\frac{Kl}{d}\right)_y = \frac{(1.0)\,(17 \times 12)\text{ in}}{6.75\text{ in}} = 30.22$$

Design Parameters

$$F_c = \frac{0.822\,E'_{y\,min}}{\left(\dfrac{Kl}{d}\right)_y^2} = \frac{0.822\,(558110\text{ psi})}{(30.22)^2} = 502.34\text{ psi}$$

$$\frac{F_{cE}}{F_c^*} = \frac{502.34\text{ psi}}{839.5\text{ psi}} = 0.598$$

Column Stability Factor, C_p $C_p = 0.5363$ (Use *Appendix, Table B-6 and Linear Interpolation*)

Allowable Compressive Stress $F_c' = F_c^* \times (C_P) = 839.5 \text{ psi} \times 0.5363 = 450.22 \text{ psi}$

Allowable Compressive Load for y-axis buckling, $P' = F_c'\,(A) = 450.22 \text{ psi} \ (74.25 \text{ in}^2) = 33429 \text{ lb}$

Allowable Compressive Load for x-axis buckling, $P' = 32999 \text{ lb}$ is less than   33429 lb.

Therefore, x-axis buckling controls the design of this glulam column and the compressive column strength is $\boxed{32999 \text{ lb.}}$

Case Study 8-3b Determine the axial compression load capacity of the Western Species glulam Column 2 in the figure sawn below. The column has a cross-section of 5 ½ x 9 and is continuous to from the ground floor to the rafter. The stress class of the gluelam members is **24F-1.8E**. MC will exceed 16 percent. Normal temperatures apply. The critical load combination has been found to be D.L + S.L.

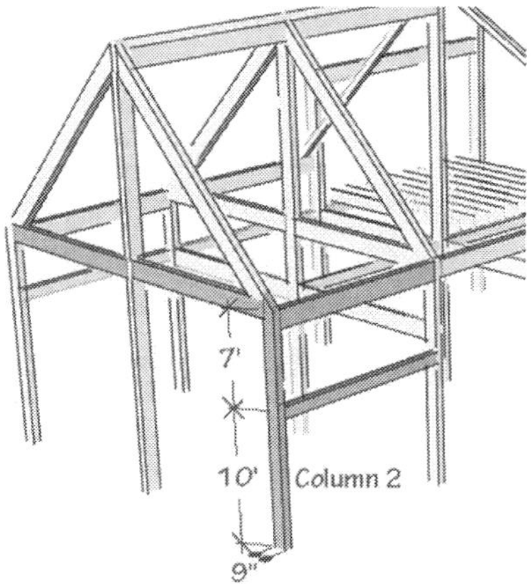

7'

10' Column 2

9"

Section Properties

$b =$ $d =$ $A =$

Adjusted Design Values

	Base Values	C_D	C_M	C_t	Allowable Values
$F_c{}^* =$					
$E'_{x\,min} =$					
$E'_{y\,min} =$					

X-axis Buckling

x-plane			
	$K_x =$	$L_x =$	$\left(\dfrac{Kl}{d}\right)_x =$

Design Parameters

$$F_c = \frac{0.822\, E'_{x\,min}}{\left(\dfrac{Kl}{d}\right)_x^2} =$$

$$\frac{F_{cE}}{F_c{}^*} =$$

Column Stability Factor, C_p $C_p =$ (Use *Appendix, Table B-6 and Linear Interpolation*)

Allowable Compressive Stress $F_c' = F_c{}^* \times (C_P) =$

Allowable Compressive Load for x-axis buckling, $P' = F_c'(A) =$

Y-axis Buckling

y-plane			
	$K_y =$	$L_y =$	$\left(\dfrac{Kl}{d}\right)_y =$

Design Parameters

$$F_c = \frac{0.822\, E'_{y\,min}}{\left(\dfrac{Kl}{d}\right)^2_y} = \qquad\qquad \frac{F_{cE}}{F_c^*} =$$

Column Stability Factor, C_p $\qquad$ $C_p =$ $\qquad\qquad$ (Use *Appendix, Table B-6 and Linear Interpolation*)

Allowable Compressive Stress $\qquad$ $F_c' = F_c^* \times (C_P) =$

Allowable Compressive Load for y-axis buckling, $P' = F_c'\,(A) =$

Allowable Compressive Load for x-axis buckling, $P' =$

Workshop 8-3a Determine the axial compression load capacity of the Western Species glulam Column 3 in the figure sawn below. The column has a cross-section of 5 ½ x 9 and is continuous to from the ground floor to the rafter. The stress class of the gluelam members is **24F-1.8E**. MC will exceed 16 percent. Normal temperatures apply. The critical load combination has been found to be D.L + S.L.

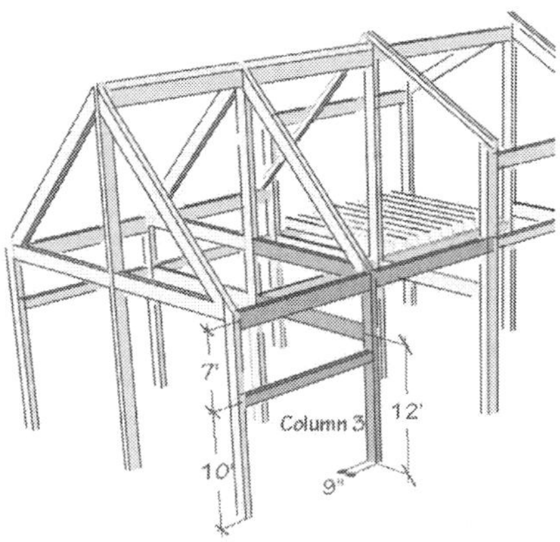

Section Properties

$b =$	$d =$	$A =$

Reference Design Values

$F_c =$	$E_{min} =$

Adjusted Design Values

	Ref. Values	C_F	C_D	C_M	C_t	C_i	Allowable Values
$F_c* =$							
$E'_{x\,min} =$		N/A	N/A			N/A	
$E'_{y\,min} =$		N/A	N/A			N/A	

X-axis Buckling

x-plane

	$K_x =$	$L_x =$	$\left(\dfrac{Kl}{d}\right)_x =$

Design Parameters

$$F_{cE} = \frac{0.822\, E'_{min}}{\left(\dfrac{Kl}{d}\right)_x^2} =$$

$$\frac{F_{cE}}{F_c^*} =$$

Column Stability Factor, C_p $C_p =$ (Use *Appendix, Table B-6 and Linear Interpolation*)

Allowable Compressive Stress $F_c' = F_c^* \times (C_P) =$

Allowable Compressive Load $P = F_c'\,(A) =$

Y-axis Buckling

y-plane	$K_y =$	$L_y =$	$\left(\dfrac{Kl}{d}\right)_y =$

Design Parameters

$$F_{cE} = \frac{0.822\, E'_{min}}{\left(\dfrac{Kl}{d}\right)_y^2} = \qquad\qquad \frac{F_{cE}}{F_c^*} =$$

Column Stability Factor, C_p $\qquad C_p =$ $\qquad\qquad$ (Use *Appendix, Table B-6 and Linear Interpolation*)

Allowable Compressive Stress $\quad F_c' = F_c^* \times (C_P) =$

Allowable Compressive Load $\qquad P = F_c'(A) =$

Workshop 8-3b Determine the axial compression load capacity of the Western Species glulam Column 4 in the figure sawn below. The column has a cross-section of 6 ¾ x 9 and is continuous to from the ground floor to the rafter. The stress class of the gluelam members is **24F-1.8E**. MC will not exceed 16 percent. Normal temperatures apply. The critical load combination has been found to be D.L + L.L.

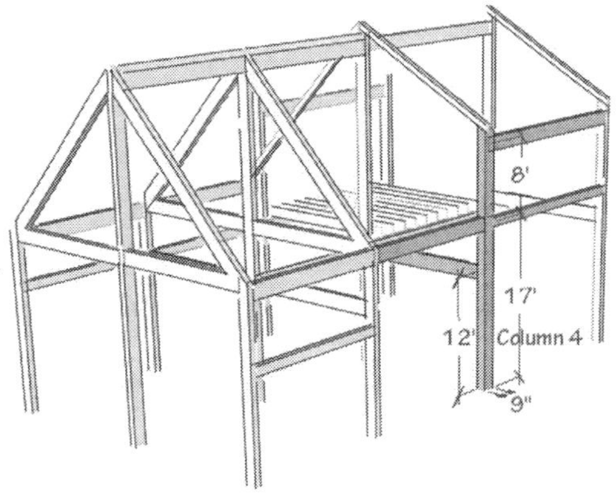

Section Properties

$b =$	$d =$	$A =$

Reference Design Values

$F_c =$	$E_{min} =$

Adjusted Design Values

	Ref. Values	C_F	C_D	C_M	C_t	C_i	Allowable Values
$F_c{}^* =$							
$E'_{x\,min} =$		N/A	N/A			N/A	
$E'_{y\,min} =$		N/A	N/A			N/A	

X-axis Buckling

x-plane

$K_x =$	$L_x =$	$\left(\dfrac{Kl}{d}\right)_x =$	

Design Parameters

$$F_{cE} = \dfrac{0.822\,E'_{min}}{\left(\dfrac{Kl}{d}\right)_x^2} =$$

$$\dfrac{F_{cE}}{F_c{}^*} =$$

Column Stability Factor, C_p $C_p =$ (Use *Appendix, Table B-6 and Linear Interpolation*)

Allowable Compressive Stress $F_c' = F_c{}^* \times (C_P) =$

Allowable Compressive Load $P = F_c'(A) =$

Y-axis Buckling

y-plane			
	$K_y =$	$L_y =$	$\left(\dfrac{Kl}{d}\right)_y =$

Design Parameters

$$F_{cE} = \frac{0.822\, E'_{min}}{\left(\dfrac{Kl}{d}\right)_y^2} =$$

$$\frac{F_{cE}}{F_c^*} =$$

Column Stability Factor, C_p $C_p =$ (Use *Appendix, Table B-6 and Linear Interpolation*)

Allowable Compressive Stress $F_c' = F_c^* \times (C_P) =$

Allowable Compressive Load $P = F_c'(A) =$

APPENDIX A

ADJUSTMENT FACTORS

Applicability of Adjustment Factors for Sawn Lumber **Table A-1**

Adjusted Design Values	Base Values	ASD only — Load duration factor	Size factor	Repetitive member factor	Temperature factor	Beam stability factor	Wet service factor	Flat use factor	Incising factor	Column stability factor	Buckling stiffness factor	Bearing area factor	Format conversion	Resistance factor	Time effect factor
$F_b{'} = F_b \times$		C_D	C_F	C_r	C_t	C_L	C_M	C_{fu}	C_i	–	–	–	K_F	ϕ_b	λ
$F_t{'} = F_t \times$		C_D	C_F	–	C_t	–	C_M	–	C_i	–	–	–	K_F	ϕ_t	λ
$F_v{'} = F_v \times$		C_D	–	–	C_t	–	C_M	–	C_i	–	–	–	K_F	ϕ_v	λ
$F_{c\perp}{'} = F_{c\perp} \times$		–	–	–	C_t	–	C_M	–	C_i	–	–	C_b	K_F	ϕ_c	λ
$F_c{'} = F_c \times$		C_D	C_F	–	C_t	–	C_M	–	C_i	C_P	–	–	K_F	ϕ_c	λ
$E{'} = E \times$		–	–	–	C_t	–	C_M	–	C_i	–	–	–	–	–	–
$E_{min}{'} = E_{min} \times$		–	–	–	C_t	–	C_M	–	C_i	–	C_T	–	K_F	ϕ_s	–

Size Factors (C_F) for Dimension Lumber **Table A-2**

Grades	Nominal Width	F_b — 2" & 3" thick	F_b — 4" thick	F_t	F_c	Other
Select Structural, No. 1 & Better, No. 1, No. 2, & No. 3	2", 3", & 4"	1.5	1.5	1.5	1.15	1.0
	5"	1.4	1.4	1.4	1.1	1.0
	6"	1.3	1.3	1.3	1.1	1.0
	8"	1.2	1.3	1.2	1.05	1.0
	10"	1.1	1.2	1.1	1.0	1.0
	12"	1.0	1.1	1.0	1.0	1.0
	14" & wider	0.9	1.0	0.9	0.9	1.0
Construction & Standard Utility	2", 3", & 4"	1.0	1.0	1.0	1.0	1.0
	2" & 3"	0.4	–	0.4	0.4	1.0
	4"	1.0	1.0	1.0	1.0	1.0
Stud	2", 3", & 4"	1.1	1.1	1.1	1.05	1.0
	5" & 6"	1.0	1.0	1.0	1.0	1.0
	8" & wider	Use No. 3 grade Reference Design Values and Size Factors				

Size Factors (C_F) for Timber — Table A-3

Nominal Depth	Net Depth	Size Factor $C_F = (12/d)^{1/9}$
12" or less	(varies)	1.0
14"	13.5"	0.987
16"	15.5"	0.972
20"	19.5"	0.947
22"	21.5"	0.937
24"	23.5"	0.928
26"	25.5"	0.920
28"	27.5"	0.912
30"	29.5"	0.905

Note: The size factor for timber is applied only to bending.

Load Duration Factors — Table A-4

Load Type	C_D	Time Frame
Permanent (Dead Load)	0.9	Greater than 10 years
Normal (**Floor Live Load**)	**1.0**	**10 years**
Snow Load	1.15	2 months
Roof Live Load	1.25	7 days
Wind or Seismic Load	1.6	10 minutes
Impact Load	2.0	Less than 2 seconds

Flat Use Factors (C_{fu}) — Table A-5

Nominal Width	Nominal Thickness	
	2" & 3"	4"
2" & 3"	1.00	N/A
4"	1.10	1.00
5"	1.10	1.05
6"	1.15	1.05
8"	1.15	1.05
10" & wider	1.20	1.10

Wet Service Factors (C_M) for Sawn Lumber — Table A-6

use with	F_b [1]	F_t	F_c [2]	F_v	$F_{c\perp}$	E & E_{min}
C_M	0.85 [1]	1.0	0.8 [2]	0.97	0.67	0.9

[1] when $(F_b \times C_F) \leq 1150$ **psi**, $C_M = 1.0$
[2] when $(F_c \times C_F) \leq 750$ **psi**, $C_M = 1.0$

Wet Service Factors (C_M) for Glulam   Table A-7

use with	F_b	F_t	F_c	F_v	$F_{c\perp}$	E & E_{min}
C_M	0.8	0.8	0.73	0.875	0.53	0.833

Temperature Factor (C_t)   Table A-8

Values	Moisture Conditions	$T \leq (100°F)$	C_t $(100°F \leq T \leq 125°F)$	$(125°F \leq T \leq 150°F)$
F_t, E, E_{min}	Wet or Dry	1.0	0.9	0.9
F_b, F_v, F_c, $F_{c\perp}$	Dry	1.0	0.8	0.7
	Wet	1.0	0.7	0.5

Bearing Area Factors (C_b)   Table A-9

L_b	0.5"	1"	1.5"	2"	3"	4"	6" or more
C_b	1.75	1.38	1.25	1.19	1.13	1.10	1.00

Format Conversion Factors (K_F)   Table A-10

Application	Property	K_F
Members	F_b, F_t, F_v, F_c, F_{rt}, F_s	$2.16/\phi$
	F_{c-p}	$1.875/\phi$
	E_{min}	$1.5/\phi$
Connections	(all connections)	$2.16/\phi$

Note: ϕ = resistance factor. K_F is not applicable where the LRFD design values are determined in accordance with the reliability normalization factor method (ASTM D 5457).

Resistance Factors (ϕ)   Table A-11

Application	Property	Symbol	Value
Members :	F_b	ϕ_b	0.85
	F_t	ϕ_t	0.80
	F_v, F_{rt}, F_s	ϕ_v	0.75
	F_c, $F_{c\perp}$	ϕ_c	0.90
	E_{min}	ϕ_s	0.85
Connections:	(all types)	ϕ_z	0.65

Time Effect Factors (λ)   Table A-12

Load Combination	λ
1.4(D+F)	0.6
1.2(D+F) + 1.6(H) + 0.5(L$_r$ or S or R)	0.6
1.2(D+F) + 1.6(L+H) + 0.5(L$_r$ or S or R)	0.7 (L is from storage)
	0.8 (L is from occupancy)
	1.25 (L is from impact)
1.2D + 1.6(L$_r$ or S or R) + (L or 0.8W)	0.8
1.2D + 1.6W + L + 0.5(L$_r$ or S or R)	1.0
1.2D + 1.0E + L + 0.2S	1.0
0.9D + 1.6W + 1.6H	1.0
0.9D + 1.0E + 1.6H	1.0

Notes:
1. Time effect factors exceeding 1.0 are not applicable to: connections, pressure borne structural members, or structural members treated with fire retardant chemicals.
2. Load combinations shown above are in accordance with ASCE 7-02. Dependent on your own design codes the above load combinations may change.

Flat Use Factors (C_{fu}) for Glulam   Table A-13

Member Dimension parallel to wide faces of laminations	C_{fu}
10-3/4" or 10-1/2"	1.01
8-3/4" or 8-1/2"	1.04
6-3/4"	1.07
5-1/8" or 5"	1.10
3-1/8" or 3"	1.16
2-1/2"	1.19

Applicability of Adjustment Factors for Glulam Timber   Table A-14

Adjusted Design Values	Base Values		ASD only — Load duration factor	ASD and LRFD — Beam stability factor	Volume factor	Temperature factor	Wet service factor	Flat use factor	Curvature factor	Column stability factor	Buckling stiffness factor	Bearing area factor	LRFD only — Format conversion	Resistance factor	Time effect factor
$F_b' = F_b$	x		C_D	C_L	C_V	C_t	C_M	C_{fu}	C_C	–	–	–	K_F	ϕ_b	λ
$F_t' = F_t$	x		C_D	–	–	C_t	C_M	–	C_C	–	–	–	K_F	ϕ_t	λ
$F_v' = F_v$	x		C_D	–	–	C_t	C_M	–	–	–	–	–	K_F	ϕ_v	λ
$F_{c\perp}' = F_{c\perp}$	x		–	–	–	C_t	C_M	–	–	–	–	C_b	K_F	ϕ_c	λ
$F_c' = F_c$	x		C_D	–	–	C_t	C_M	–	–	C_P	–	–	K_F	ϕ_c	λ
$E' = E$	x		–	–	–	C_t	C_M	–	–	–	–	–	–	–	–
$E_{min}' = E_{min}$	x		–	–	–	C_t	C_M	–	–	–	C_T	–	K_F	ϕ_s	–

APPENDIX

B

REFERENCE DESIGN VALUES & SECTION PROPERTIES

Reference Design Values for Visually Graded Dimension Lumber (psi) Table B-1

Species or Group	Grade	Extreme Fiber Stress in Bending F_b	Tension Parallel to Grain F_t	Horizontal Shear F_v	Compression Perpendicular $F_{c\perp}$	Compression Parallel to Grain F_c	Modulus of Elasticity E	Modulus of Elasticity E_{min}
Douglas Fir-Larch Douglas Fir Western Larch	Select Structural	1500	1000	180	625	1700	1,900,000	690,000
	No. 1 & Btr.	1200	800	180	625	1550	1,800,000	660,000
	No. 1	1000	675	180	625	1500	1,700,000	620,000
	No. 2	900	575	180	625	1350	1,600,000	580,000
	No. 3	525	325	180	625	775	1,400,000	510,000
	Construction	1000	650	180	625	1650	1,500,000	550,000
	Standard	575	375	180	625	1400	1,400,000	510,000
	Utility	275	175	180	625	900	1,300,000	470,000
	Stud	700	450	180	625	850	1,400,000	510,000
Douglas Fir-South Douglas Fir-South (grown in AZ, CO, NV, NM and UT)	Select Structural	1350	900	180	520	1600	1,400,000	510,000
	No. 1	925	600	180	520	1450	1,300,000	470,000
	No. 2	850	525	180	520	1350	1,200,000	440,000
	No. 3	500	300	180	520	775	1,100,000	400,000
	Construction	975	600	180	520	1650	1,200,000	440,000
	Standard	550	350	180	520	1400	1,100,000	400,000
	Utility	250	150	180	520	900	1,000,000	370,000
	Stud	675	425	180	520	850	1,100,000	400,000
Hem-Fir Western Hemlock Noble Fir California Red Fir Grand Fir Pacific Silver Fir White Fir	Select Structural	1400	925	150	405	1500	1,600,000	580,000
	No. 1 & Btr.	1100	725	150	405	1350	1,500,000	550,000
	No. 1	975	625	150	405	1350	1,500,000	550,000
	No. 2	850	525	150	405	1300	1,300,000	470,000
	No. 3	500	300	150	405	725	1,200,000	440,000
	Construction	975	600	150	405	1550	1,300,000	470,000
	Standard	550	325	150	405	1300	1,200,000	440,000
	Utility	250	150	150	405	850	1,100,000	400,000
	Stud	675	400	150	405	800	1,200,000	440,000
Spruce-Pine-Fir (South) Western Species: Engelmann Spruce Sitka Spruce White Spruce Lodgepole Pine	Select Structural	1300	575	135	335	1200	1,300,000	470,000
	No. 1	875	400	135	335	1050	1,200,000	440,000
	No. 2	775	350	135	335	1000	1,100,000	400,000
	No. 3	450	200	135	335	575	1,000,000	370,000
	Construction	875	400	135	335	1200	1,000,000	370,000
	Standard	500	225	135	335	1000	900,000	330,000
	Utility	225	100	135	335	675	900,000	330,000
	Stud	600	275	135	335	625	1,000,000	370,000
Western Cedars Western Red Cedar Incense Cedar Port Orford Cedar Alaskan Cedar	Select Structural	1000	600	155	425	1000	1,100,000	400,000
	No. 1	725	425	155	425	825	1,000,000	370,000
	No. 2	700	425	155	425	650	1,000,000	370,000
	No. 3	400	250	155	425	375	900,000	330,000
	Construction	800	475	155	425	850	900,000	330,000
	Standard	450	275	155	425	650	800,000	290,000
	Utility	225	125	155	425	425	800,000	290,000
	Stud	550	325	155	425	400	900,000	330,000
Western Woods) (and White Woods Idaho White Pine Ponderosa Pine, Sugar Pine Alpine Fir Mountain Hemlock	Select Structural	900	400	135	335	1050	1,200,000	440,000
	No. 1	675	300	135	335	950	1,100,000	400,000
	No. 2	675	300	135	335	900	1,000,000	370,000
	No. 3	375	175	135	335	525	900,000	330,000
	Construction	775	350	135	335	1100	1,000,000	370,000
	Standard	425	200	135	335	925	900,000	330,000
	Utility	200	100	135	335	600	800,000	290,000
	Stud	525	225	135	335	575	900,000	330,000

Reference Design Values for Beams & Stringers (psi)   **Table B-2**
5" and thicker, width more than 2" greater than thickness

Species or Group	Grade	Extreme Fiber Stress in Bending F_b	Tension Parallel to Grain F_t	Horizontal Shear F_v	Compression Perpen-dicular $F_{c\perp}$	Compression Parallel to Grain F_c	Modulus of Elasticity E	Modulus of Elasticity E_{min}
Douglas Fir-Larch	Dense Select Structural	1750	1150	170	730	1350	1,700,000	620,000
	Dense No. 1	1400	950	170	730	1200	1,700,000	620,000
	Dense No. 2	850	550	170	730	825	1,400,000	510,000
	Select Structural	1500	1000	170	625	1150	1,600,000	580,000
	No. 1	1200	825	170	625	1000	1,600,000	580,000
	No. 2	750	475	170	625	700	1,300,000	470,000
Douglas Fir-South	Select Structural	1450	950	165	520	1050	1,200,000	440,000
	No. 1	1150	775	165	520	925	1,200,000	440,000
	No. 2	675	450	165	520	650	1,000,000	370,000
Hem-Fir	Select Structural	1200	800	140	405	975	1,300,000	470,000
	No. 1	975	650	140	405	850	1,300,000	470,000
	No. 2	575	375	140	405	575	1,100,000	400,000
Mountain Hemlock	Select Structural	1250	825	170	570	925	1,100,000	400,000
	No. 1	1000	675	170	570	800	1,100,000	400,000
	No. 2	625	400	170	570	550	900,000	330,000
Sitka Spruce	Select Structural	1150	750	140	435	875	1,300,000	470,000
	No. 1	925	600	140	435	750	1,300,000	470,000
	No. 2	550	350	140	435	525	1,100,000	400,000
Spruce-Pine-Fir (South)	Select Structural	1000	675	125	335	700	1,200,000	440,000
	No. 1	875	550	125	335	625	1,200,000	440,000
	No. 2	475	325	125	335	425	1,000,000	370,000
Western Cedars	Select Structural	1100	725	140	425	925	1,000,000	370,000
	No. 1	875	600	140	425	800	1,000,000	370,000
	No. 2	550	350	140	425	550	800,000	290,000
Western Hemlock	Select Structural	1300	875	170	410	1100	1,400,000	510,000
	No. 1	1050	700	170	410	950	1,400,000	510,000
	No. 2	650	425	170	410	650	1,100,000	400,000
Western Woods (and White Woods)	Select Structural	1000	675	125	345	800	1,100,000	400,000
	No. 1	800	525	125	345	700	1,100,000	400,000
	No. 2	475	325	125	345	475	900,000	330,000

Reference Design Values for Posts & Timbers (psi)   Table B-3
5"x5" and larger, width not more than 2" greater than thickness

Species or Group	Grade	Extreme Fiber Stress in Bending F_b	Tension Parallel to Grain F_t	Horizontal Shear F_v	Compression Perpen-dicular $F_{c\perp}$	Compression Parallel to Grain F_c	Modulus of Elasticity E	Modulus of Elasticity E_{min}
Douglas Fir-Larch	Dense Select Structural	1750	1150	170	730	1350	1,700,000	620,000
	Dense No. 1	1400	950	170	730	1200	1,700,000	620,000
	Dense No. 2	850	550	170	730	825	1,400,000	510,000
	Select Structural	1500	1000	170	625	1150	1,600,000	580,000
	No. 1	1200	825	170	625	1000	1,600,000	580,000
	No. 2	750	475	170	625	700	1,300,000	470,000
Douglas Fir-South	Select Structural	1450	950	165	520	1050	1,200,000	440,000
	No. 1	1150	775	165	520	925	1,200,000	440,000
	No. 2	675	450	165	520	650	1,000,000	370,000
Hem-Fir	Select Structural	1200	800	140	405	975	1,300,000	470,000
	No. 1	975	650	140	405	850	1,300,000	470,000
	No. 2	575	375	140	405	575	1,100,000	400,000
Mountain Hemlock	Select Structural	1250	825	170	570	925	1,100,000	400,000
	No. 1	1000	675	170	570	800	1,100,000	400,000
	No. 2	625	400	170	570	550	900,000	330,000
Sitka Spruce	Select Structural	1150	750	140	435	875	1,300,000	470,000
	No. 1	925	600	140	435	750	1,300,000	470,000
	No. 2	550	350	140	435	525	1,100,000	400,000
Spruce-Pine-Fir (South)	Select Structural	1000	675	125	335	700	1,200,000	440,000
	No. 1	875	550	125	335	625	1,200,000	440,000
	No. 2	475	325	125	335	425	1,000,000	370,000
Western Cedars	Select Structural	1100	725	140	425	925	1,000,000	370,000
	No. 1	875	600	140	425	800	1,000,000	370,000
	No. 2	550	350	140	425	550	800,000	290,000
Western Hemlock	Select Structural	1300	875	170	410	1100	1,400,000	510,000
	No. 1	1050	700	170	410	950	1,400,000	510,000
	No. 2	650	425	170	410	650	1,100,000	400,000
Western Woods (and White Woods)	Select Structural	1000	675	125	345	800	1,100,000	400,000
	No. 1	800	525	125	345	700	1,100,000	400,000
	No. 2	475	325	125	345	475	900,000	330,000

Reference Design Values for Glulam

Table B-4

Stress Class	Bending About X-X Axis (Loaded Perpendicular to Wide Faces of Laminations)						Bending About Y-Y Axis (Loaded Parallel to Wide Faces of Laminations)					Axially Loaded			Fasteners
	Extreme Fiber in Bending		Compression Perpendicular to Grain	Shear Parallel to Grain (Horizontal)	Modulus of Elasticity	Modulus of Elasticity for Beam and Column Stability	Extreme Fiber in Bending	Compression Perpendicular to Grain	Shear Parallel to Grain (Horizontal)	Modulus of Elasticity	Modulus of Elasticity for Beam and Column Stability	Tension Parallel to Grain	Compression Parallel to Grain	Modulus of Elasticity	Specific Gravity for Fastener Design
	Tension Zone Stressed in Tension (Positive Bending)	Compression Zone Stressed in Tension (Negative Bending)													
	F_{bx}^{+} (psi)	$F_{bx}^{-(1)}$ (psi)	$F_{c\perp x}$ (psi)	$F_{vx}^{(4)}$ (psi)	E_x (10^6 psi)	$E_{x\,min}$ (10^6 psi)	F_{by} (psi)	$F_{c\perp y}$ (psi)	$F_{vy}^{(4)(5)}$ (psi)	E_y (10^6 psi)	$E_{y\,min}$ (10^6 psi)	F_t (psi)	F_c (psi)	E_{axial} (10^6 psi)	G
16F-1.3E	1600	925	315	195	1.3	0.67	800	315	170	1.1	0.57	675	925	1.2	0.42
20F-1.5E	2000	1100	425	210 [6]	1.5	0.78	800	315	185	1.2	0.62	725	925	1.3	0.42
24F-1.7E	2400	1450	500	210	1.7	0.88	1050	315	185	1.3	0.67	775	1000	1.4	0.42
24F-1.8E	2400	1450 [2]	650	265 [3]	1.8	0.93	1450	560	230 [3]	1.6	0.83	1100	1600	1.7	0.50
26F-1.9E [7]	2600	1950	650	265 [3]	1.9	0.98	1600	560	230 [3]	1.6	0.83	1150	1600	1.7	0.50 [10]
28F-2.1E SP [7]	2800	2300	740	300	2.1 [9]	1.09 [9]	1600	650	260	1.7	0.88	1250	1750	1.7	0.55
30F-2.1E SP [7][8]	3000	2400	740	300	2.1 [9]	1.09 [9]	1750	650	260	1.7	0.88	1250	1750	1.7	0.55

1. For balanced layups, F_{bx}^{-} shall be equal to F_{bx}^{+} for the stress class. Designer shall specify when balanced layup is required.
2. Negative bending stress, F_{bx}^{-}, is permitted to be increased to 1,850 psi for Douglas Fir and to 1,950 psi for Southern Pine for specific combinations. Designer shall specify when these increased stresses are required.
3. For structural glued laminated timber of Southern Pine, the basic shear design values, F_{vx} and F_{vy}, are permitted to be increased to 300 psi and 260 psi, respectively.
4. The design value for shear, F_{vx} and F_{vy}, shall be decreased by multiplying by a factor of 0.72 for non-prismatic members, notched members, and for all members subject to impact or cyclic loading. The reduced design value shall be used for design of members at connections that transfer shear by mechanical fasteners (NDS 3.4.3.3). The reduced design value shall also be used for determination of design values for radial tension (NDS 5.2.2).
5. Design values are for timbers with laminations made from a single piece of lumber across the width or multiple pieces that have been edge bonded. For timbers manufactured from multiple piece laminations (across width) that are not edge bonded, value shall be multiplied by 0.4 for members with 5, 7, or 9 laminations or by 0.5 for all other members. This reduction shall be cumulative with the adjustment in footnote (4).
6. Certain Southern Pine combinations may contain lumber with wane. If lumber with wane is used, the design value for shear parallel to grain, F_{vx}, shall be multiplied by 0.67 if wane is allowed on both sides. If wane is limited to one side, F_{vx} shall be multiplied by 0.83. This reduction shall be cumulative with the adjustment in footnote (4).
7. 26F, 28F, and 30F beams are not produced by all manufacturers; therefore, availability may be limited. Contact supplier or manufacturer for details.
8. 30F combinations are restricted to a maximum 6" nominal width.
9. For 28F and 30F members with more than 15 laminations, E_x = 2.0 million psi and $E_{x\,min}$ = 1.04 million psi.
10. For structural glued laminated timber of Southern Pine, specific gravity for fastener design is permitted to be increased to 0.55.

Design values in this table represent design values for groups of similar structural glued laminated timber combinations. Higher design values for some properties may be obtained by specifying a particular combination listed in Table 5A Expanded. Design values are for members with 4 or more laminations. For 2 and 3 lamination members, see Table 5B. Some stress classes are not available in all species. Contact structural glued laminated timber manufacturer for availability.

Section Properties of Sawn Lumber

Nominal Size (in.)	Actual Surfaced Size (in.)	Area (in²)	Section Modulus S_x (in³)	Moment of Inertia I_x (in⁴)	Section Modulus S_y (in³)	Moment of Inertia I_y (in⁴)	Board Feet (per ft.)
2 × 2	1.5 × 1.5	2.25	0.56	0.42	0.56	0.42	0.33
2 × 3	1.5 × 2.5	3.75	1.56	1.95	0.94	0.70	0.50
2 × 4	1.5 × 3.5	5.25	3.06	5.36	1.31	0.98	0.67
2 × 6	1.5 × 5.5	8.25	7.56	20.80	2.06	1.55	1.00
2 × 8	1.5 × 7.25	10.88	13.14	47.63	2.72	2.04	1.33
2 × 10	1.5 × 9.25	13.88	21.39	98.93	3.47	2.60	1.67
2 × 12	1.5 × 11.25	16.88	31.64	177.98	4.22	3.16	2.00
2 × 14	1.5 × 13.25	19.88	43.89	290.78	4.97	3.73	2.33
3 × 3	2.5 × 2.5	6.25	2.60	3.26	2.60	3.26	0.75
3 × 4	2.5 × 3.5	8.75	5.10	8.93	3.65	4.56	1.00
3 × 6	2.5 × 5.5	13.75	12.60	34.66	5.73	7.16	1.50
3 × 8	2.5 × 7.25	18.12	21.90	79.39	7.55	9.44	2.00
3 × 10	2.5 × 9.25	23.12	35.65	164.89	9.64	12.04	2.50
3 × 12	2.5 × 11.25	28.12	52.73	296.63	11.72	14.65	3.00
3 × 14	2.5 × 13.25	33.12	73.15	484.63	13.80	17.25	3.50
3 × 16	2.5 × 15.25	38.12	96.90	738.87	15.89	19.86	4.00
4 × 4	3.5 × 3.5	12.25	7.15	12.51	7.15	12.51	1.33
4 × 6	3.5 × 5.5	19.25	17.65	48.53	11.23	19.65	2.00
4 × 8	3.5 × 7.25	25.38	30.66	111.15	14.80	25.90	2.67
4 × 10	3.5 × 9.25	32.38	49.91	230.84	18.89	33.05	3.33
4 × 12	3.5 × 11.25	39.38	73.83	415.28	22.97	40.20	4.00
4 × 14	3.5 × 13.25	46.38	102.41	678.48	27.05	47.34	4.67
4 × 16	3.5 × 15.25	53.38	135.66	1034.42	31.14	54.49	5.33
6 × 6	5.5 × 5.5	30.25	27.73	76.26	27.73	76.26	3.00
6 × 8	5.5 × 7.5	41.25	51.56	193.36	37.81	103.98	4.00
6 × 10	5.5 × 9.5	52.25	82.73	392.96	47.90	131.71	5.00
6 × 12	5.5 × 11.5	63.25	121.23	697.07	57.98	159.44	6.00
6 × 14	5.5 × 13.5	74.25	167.06	1127.67	68.06	187.17	7.00
6 × 16	5.5 × 15.5	85.25	220.23	1706.78	78.15	214.90	8.00
6 × 18	5.5 × 17.5	96.25	280.73	2456.38	88.23	242.63	9.00
6 × 20	5.5 × 19.5	107.25	348.56	3398.48	98.31	270.36	10.00
8 × 8	7.5 × 7.5	56.25	70.31	263.67	70.31	263.67	5.33
8 × 10	7.5 × 9.5	71.25	112.81	535.86	89.06	333.98	6.67
8 × 12	7.5 × 11.5	86.25	165.31	950.55	107.81	404.30	8.00
8 × 14	7.5 × 13.5	101.25	227.81	1537.73	126.56	474.61	9.33
8 × 16	7.5 × 15.5	116.25	300.31	2327.42	145.31	544.92	10.67
8 × 18	7.5 × 17.5	131.25	382.81	3349.61	164.06	615.23	12.00
8 × 20	7.5 × 19.5	146.25	475.31	4634.30	182.81	685.55	13.33
8 × 22	7.5 × 21.5	161.25	577.81	6211.48	201.56	755.86	14.67
8 × 24	7.5 × 23.5	176.25	690.31	8111.17	220.31	826.17	16.00
10 × 10	9.5 × 9.5	90.25	142.90	678.76	142.90	678.76	8.33
10 × 12	9.5 × 11.5	109.25	209.40	1204.03	172.98	821.65	10.00
10 × 14	9.5 × 13.5	128.25	288.56	1947.80	203.06	964.55	11.67
10 × 16	9.5 × 15.5	147.25	380.40	2948.07	233.15	1107.44	13.33
10 × 18	9.5 × 17.5	166.25	484.90	4242.84	263.23	1250.34	15.00
10 × 20	9.5 × 19.5	185.25	602.06	5870.11	293.31	1393.23	16.67
10 × 22	9.5 × 21.5	204.25	731.90	7867.88	323.40	1536.13	18.33
12 × 12	11.5 × 11.5	132.25	253.48	1457.51	253.48	1457.51	12.00
12 × 14	11.5 × 13.5	155.25	349.31	2357.86	297.56	1710.98	14.00
12 × 16	11.5 × 15.5	178.25	460.48	3568.71	341.65	1964.46	16.00

Column Stability Factors, C_P

Table B-6

F_{CE}/F_c^*	Sawn C_P	Glulam C_P	F_{CE}/F_c^*	Sawn C_P	Glulam C_P	F_{CE}/F_c^*	Sawn C_P	Glulam C_P	F_{CE}/F_c^*	Sawn C_P	Glulam C_P
0.00	0.0000	0.0000	0.61	0.5062	0.5448	1.24	0.7597	0.8311	2.55	0.9014	0.9444
0.01	0.0100	0.0100	0.62	0.5123	0.5520	1.26	0.7643	0.8355	2.60	0.9037	0.9459
0.02	0.0199	0.0200	0.63	0.5184	0.5591	1.28	0.7688	0.8398	2.65	0.9059	0.9473
0.03	0.0298	0.0299	0.64	0.5244	0.5661	1.30	0.7731	0.8439	2.70	0.9080	0.9486
0.04	0.0397	0.0398	0.65	0.5303	0.5731	1.32	0.7773	0.8478	2.75	0.9100	0.9499
0.05	0.0495	0.0497	0.66	0.5361	0.5799	1.34	0.7814	0.8515	2.80	0.9119	0.9511
0.06	0.0593	0.0596	0.67	0.5418	0.5867	1.36	0.7853	0.8551	2.85	0.9138	0.9522
0.07	0.0690	0.0695	0.68	0.5475	0.5934	1.38	0.7892	0.8586	2.90	0.9155	0.9533
0.08	0.0787	0.0793	0.69	0.5531	0.6000	1.40	0.7929	0.8619	2.95	0.9172	0.9544
0.09	0.0883	0.0891	0.70	0.5586	0.6065	1.42	0.7965	0.8651	3.00	0.9189	0.9554
0.10	0.0979	0.0989	0.71	0.5640	0.6129	1.44	0.8000	0.8682	3.05	0.9204	0.9563
0.11	0.1074	0.1087	0.72	0.5694	0.6193	1.46	0.8034	0.8711	3.10	0.9219	0.9572
0.12	0.1169	0.1184	0.73	0.5747	0.6255	1.48	0.8067	0.8740	3.15	0.9234	0.9581
0.13	0.1263	0.1281	0.74	0.5799	0.6317	1.50	0.8099	0.8767	3.20	0.9248	0.9590
0.14	0.1357	0.1378	0.75	0.5850	0.6377	1.52	0.8130	0.8793	3.25	0.9262	0.9598
0.15	0.1451	0.1474	0.76	0.5901	0.6437	1.54	0.8160	0.8818	3.30	0.9275	0.9606
0.16	0.1544	0.1571	0.77	0.5951	0.6496	1.56	0.8190	0.8843	3.35	0.9287	0.9613
0.17	0.1636	0.1667	0.78	0.6000	0.6554	1.58	0.8218	0.8866	3.40	0.9300	0.9620
0.18	0.1728	0.1762	0.79	0.6048	0.6611	1.60	0.8246	0.8889	3.45	0.9312	0.9627
0.19	0.1819	0.1858	0.80	0.6096	0.6667	1.62	0.8273	0.8911	3.50	0.9323	0.9634
0.20	0.1910	0.1953	0.81	0.6143	0.6722	1.64	0.8299	0.8932	3.55	0.9334	0.9641
0.21	0.2000	0.2047	0.82	0.6189	0.6776	1.66	0.8325	0.8952	3.60	0.9345	0.9647
0.22	0.2090	0.2142	0.83	0.6235	0.6829	1.68	0.8350	0.8972	3.65	0.9355	0.9653
0.23	0.2179	0.2236	0.84	0.6280	0.6881	1.70	0.8374	0.8991	3.70	0.9365	0.9659
0.24	0.2267	0.2329	0.85	0.6324	0.6933	1.72	0.8398	0.9009	3.75	0.9375	0.9664
0.25	0.2355	0.2423	0.86	0.6368	0.6983	1.74	0.8421	0.9027	3.80	0.9384	0.9670
0.26	0.2442	0.2515	0.87	0.6410	0.7033	1.76	0.8443	0.9044	3.85	0.9394	0.9675
0.27	0.2529	0.2608	0.88	0.6453	0.7082	1.78	0.8465	0.9061	3.90	0.9403	0.9680
0.28	0.2615	0.2700	0.89	0.6494	0.7129	1.80	0.8486	0.9077	3.95	0.9411	0.9685
0.29	0.2700	0.2792	0.90	0.6535	0.7176	1.82	0.8507	0.9092	4.00	0.9420	0.9690
0.30	0.2785	0.2883	0.91	0.6575	0.7222	1.84	0.8527	0.9107	4.05	0.9428	0.9695
0.31	0.2869	0.2974	0.92	0.6615	0.7267	1.86	0.8547	0.9122	4.10	0.9436	0.9699
0.32	0.2953	0.3065	0.93	0.6654	0.7312	1.88	0.8566	0.9136	4.15	0.9444	0.9704
0.33	0.3035	0.3155	0.94	0.6692	0.7355	1.90	0.8585	0.9150	4.20	0.9451	0.9708
0.34	0.3118	0.3244	0.95	0.6730	0.7397	1.92	0.8603	0.9163	4.25	0.9458	0.9712
0.35	0.3199	0.3333	0.96	0.6767	0.7439	1.94	0.8621	0.9176	4.30	0.9466	0.9716
0.36	0.3280	0.3422	0.97	0.6804	0.7480	1.96	0.8638	0.9189	4.35	0.9473	0.9720
0.37	0.3360	0.3510	0.98	0.6840	0.7520	1.98	0.8656	0.9201	4.40	0.9479	0.9724
0.38	0.3439	0.3598	0.99	0.6875	0.7559	2.00	0.8672	0.9213	4.45	0.9486	0.9728
0.39	0.3518	0.3685	1.00	0.6910	0.7597	2.02	0.8688	0.9225	4.50	0.9492	0.9731
0.40	0.3596	0.3772	1.01	0.6944	0.7635	2.04	0.8704	0.9236	4.55	0.9499	0.9735
0.41	0.3673	0.3858	1.02	0.6978	0.7672	2.06	0.8720	0.9247	4.60	0.9505	0.9738
0.42	0.3750	0.3943	1.03	0.7011	0.7708	2.08	0.8735	0.9258	4.65	0.9511	0.9742
0.43	0.3826	0.4028	1.04	0.7044	0.7743	2.10	0.8750	0.9268	4.70	0.9517	0.9745
0.44	0.3901	0.4113	1.05	0.7076	0.7778	2.12	0.8765	0.9278	4.75	0.9522	0.9748
0.45	0.3975	0.4197	1.06	0.7107	0.7812	2.14	0.8779	0.9288	4.80	0.9528	0.9751
0.46	0.4049	0.4280	1.07	0.7138	0.7845	2.16	0.8793	0.9297	4.85	0.9534	0.9754
0.47	0.4122	0.4362	1.08	0.7169	0.7877	2.18	0.8806	0.9307	4.90	0.9539	0.9757
0.48	0.4194	0.4444	1.09	0.7199	0.7909	2.20	0.8820	0.9316	5.00	0.9549	0.9763
0.49	0.4265	0.4526	1.10	0.7229	0.7940	2.22	0.8833	0.9325	6.00	0.9632	0.9808
0.50	0.4336	0.4607	1.11	0.7258	0.7970	2.24	0.8846	0.9333	8.00	0.9731	0.9861
0.51	0.4406	0.4687	1.12	0.7287	0.8000	2.26	0.8858	0.9342	10.00	0.9788	0.9891
0.52	0.4475	0.4766	1.13	0.7315	0.8029	2.28	0.8870	0.9350	20.00	0.9897	0.9948
0.53	0.4543	0.4845	1.14	0.7343	0.8058	2.30	0.8882	0.9358	40.00	0.9949	0.9974
0.54	0.4611	0.4923	1.15	0.7370	0.8085	2.32	0.8894	0.9366	60.00	0.9966	0.9983
0.55	0.4678	0.5000	1.16	0.7397	0.8113	2.34	0.8906	0.9374	100.00	0.9980	0.9990
0.56	0.4744	0.5077	1.17	0.7423	0.8139	2.36	0.8917	0.9381	200.00	0.9990	0.9995
0.57	0.4809	0.5152	1.18	0.7449	0.8165	2.38	0.8928	0.9388			
0.58	0.4873	0.5227	1.19	0.7475	0.8191	2.40	0.8939	0.9396			
0.59	0.4937	0.5302	1.20	0.7500	0.8216	2.45	0.8965	0.9413			
0.60	0.5000	0.5375	1.22	0.7549	0.8264	2.50	0.8990	0.9429			

INDEX